GovCloud: Cloud Computing for the Business of Government

A Five-Step Process to Evaluate, Design and Implement A Robust Cloud Solution

The Essential Desk Reference and Guide for Managers

By

Kevin Jackson

and

Don Philpott

Published by
Government Training Inc.™
ISBN: 978-0-9832361-3-9

About the Publisher – Government Training Inc. ™

Government Training Inc. provides worldwide training, publishing, and consulting to government agencies and contractors that support government in areas of business and financial management, acquisition and contracting, physical and cyber security, and intelligence operations. Our management team and instructors are seasoned executives with demonstrated experience in areas of Federal, State, Local and DoD needs and mandates.

Recent books published by Government Training Inc. ™ include:

- ☐ The COTR Handbook
- ☐ Performance Based Contracting Handbook
- ☐ Cost Reimbursable Contracting
- ☐ Handbook for Managing Teleworkers
- ☐ Handbook for Managing Teleworkers: Toolkit
- ☐ Small Business Guide to Government Contracting
- ☐ Securing Our Schools
- ☐ Workplace Violence
- ☐ The Grant Writer's Handbook
- ☐ The Integrated Physical Security Handbook

For more information on the company, its publications and professional training, go to www.GovernmentTrainingInc.com.

For information regarding permissions, write to:
Government Training Inc. ™
Rights and Contracts Department
5372 Sandhamn Place
Longboat Key, Florida 34228
don.dickson@GovernmentTrainingInc.com

ISBN: 978-0-9832361-3-9
www.GovernmentTrainingInc.com

Sources:

This book has drawn heavily on the authoritative materials published by a wide range of sources.

These materials are in the public domain, but accreditation has been given both in the text and in the reference section.

The author and publisher have taken great care in the preparation of this handbook, but make no expressed or implied warranty of any kind, and assume no responsibility for errors or omissions.

No liability is assumed for incidental or consequential damages in connection with, or arising out of, the use of the information or recommendations contained herein.

Contents

About the authors

Kevin Jackson

Mr. Kevin Jackson is currently Director, Cloud Services at NJVC, one of the largest information technology solutions providers supporting the U.S. Department of Defense (DoD). Prior to this position, he served in various senior management positions including Vice President JP Morgan Chase, Federal Services Director for Sirius Computer Solutions and Worldwide Sales Executive (Mobile and Wireless Technologies) for IBM. His formal education includes MSEE (Computer Engineering), MA National Security & Strategic Studies and a BS in Aerospace Engineering. Mr. Jackson retired from the U.S. Navy earning specialties in Space Systems Engineering, Airborne Logistics and Airborne Command and Control. He also served with the National Reconnaissance Office, Operational Support Office, providing tactical support to Navy and Marine Corps forces worldwide. Recently, Mr. Jackson has been deeply involved in the broad collaborative effort between industry and the U.S. National Institute of Standards and Technology (NIST) on the Federal adoption of cloud computing technologies. Kevin Jackson is the founder and author of "Cloud Musings" (http://kevinljackson.blogspot.com), a widely followed blog that focuses on the use of cloud computing by the Federal government. He is also the editor and founder of "Government Cloud Computing on Ulitzer" electronic magazine (http://govcloud.ulitzer.com).

Don Philpott

Don Philpott is editor of International Homeland Security, a quarterly journal for homeland security professionals, and has been writing, reporting and broadcasting on international events, trouble spots and major news stories for more than 40 years. For 20 years he was a senior correspondent with Press Association -Reuters, the wire service, and traveled the world on assignments including Northern Ireland, Lebanon, Israel, South Africa and Asia.

He writes for magazines and newspapers in the United States and Europe and is a contributor to radio and television programs on security and other issues. He is the author of more than 90 books on a wide range of subjects and has had more than 5,000 articles printed in publications around the world. His most recent books are Handbook for COTRs, Performance Based Contracting, Cost Reimbursable Contracting, How to Manage Teleworkers and just released, How to Manage Teleworkers: Toolkit. All of these books have been published by Government Training Inc.

He is a member of the National Press Club.

Symbols

Throughout this book you will see a number of icons displayed in the margins. The icons are there to help you as you work through the Five Step process. Each icon acts as an advisory – for instance, alerting you to things that you must always do or should never do. The icons used are:

Must Do This is something that you must always do

No No This is something you should never do

Tips Really useful tips

Remember Points to bear in mind

Checklist Have you checked off or answered everything on this list?

INTRODUCTION

IT and the Federal Government

Information technology should enable government to better serve the American people. But, despite spending more than $600 billion on information technology over the past decade, the Federal Government has achieved little of the productivity improvements that private industry has realized from IT. Too often, Federal IT projects run over budget, behind schedule, or fail to deliver promised functionality. Many projects use "grand design" approaches that aim to deliver functionality every few years, rather than breaking projects into more manageable chunks and demanding new functionality every few quarters. In addition, the Federal Government too often relies on large, custom, proprietary systems when "light technologies" or shared services exist.

Government officials have been trying to adopt best practices for years – from the Raines Rules of the 1990s through the Clinger Cohen Act and the acquisition regulations that followed. But obstacles have always gotten in the way. This plan attempts to clear these obstacles, allowing agencies to leverage information technology to create a more efficient and effective government.

Over the last 18 months, we have engaged the Federal IT, acquisition, and program management communities, industry experts, and academics. We have conducted listening sessions with Congress, agency CIOs, and senior procurement executives. We have received detailed input and recommendations from many industry groups such as TechAmerica. This engagement process has led to recommendations for IT reform in the areas of operational efficiency and large-scale IT program management.

A 25-point action plan has been designed to deliver more value to the American taxpayer. These actions have been planned to take place over the next 18 months and place ownership with OMB and agency operational centers, as appropriate. While the 25 points may not solve all Federal IT challenges, they will address many of the most pressing, persistent challenges. This plan requires a focus on execution and is designed to establish some early wins to garner momentum for our continued efforts. Active involvement from agency leadership is critical to the success of these reforms. As such, the Federal CIO will work with the President's Management Council to successfully implement this plan.

Some highlights of the implementation plan include:

- ☐ Turn around or terminate at least one-third of underperforming projects in IT portfolio within the next 18 months
- ☐ Shift to "Cloud First" policy. Each agency will identify three "must move" services within three months, move one of those services to the cloud within 12 months, and the remaining two within 18 months.
- ☐ Reduce the number of Federal data centers by at least 800 by the year 2015
- ☐ Only approve funding of major IT programs that:
  - Have a dedicated program manager and a fully-staffed integrated program team
  - Use a modular approach with usable functionality delivered every six months
  - Use specialized IT acquisition professionals
- ☐ Work with Congress to:
  - Consolidate commodity IT funding under the agency CIOs and
  - Develop flexible budget models that align with modular development
- ☐ Launch an interactive platform for pre-RFP agency-industry collaboration

Vivek Kundra, U.S. Chief Information Officer, The White House, A full transcript of the action plan is attached as Appendix 4.

U.S. Government IT Today

Remember

The United States Government is the world's largest consumer of information technology, spending over $76 billion annually on more than 10,000 different systems. Fragmentation of systems, poor project execution, and the drag of legacy technology in the Federal Government have presented barriers to achieving the productivity and performance gains found when technology is deployed effectively in the private sectors.

"The Obama Administration is changing the way business is done in Washington and bringing a new sense of responsibility to how we manage taxpayer dollars. We are working to bring the spirit of American innovation and the power of technology to improve performance and lower the cost of government operations," said Federal Chief Information Officer, Vivek Kundra.

In September 2009, the Federal Government announced its Cloud Computing Initiative. Cloud computing has the potential to greatly reduce waste, increase data center efficiency and utilization rates, and lower operating costs. The initiative included details on deployment models, service models, and common characteristics of cloud computing.

"As we move to the cloud, we must be vigilant in our efforts to ensure that the standards are in place for a cloud computing environment that provides for security of government information, protects the privacy of our citizens, and safeguards our national security interests," Kundra said.

Times are Changing

For the first time in memory we have three ingredients in place that are essential for a step function improvement in Federal performance, Jeffrey Zients, Federal Chief Performance Officer, told the Center for American Progress in Washington, D.C., in February 2010.

"First, we have a president who is committed to opening government—to right answers wherever they come from. I can tell you from my private sector experience that this type of openness leads to innovation and improvement. Second, the president has refrained from wholesale government bashing. While it can be appealing in its simplicity, it's counterproductive. To get real results we need to engage conscientious, hard-working people in the effort. And the president's tone paves the way.

"Finally, we have the urgency of the moment. Mounting deficits and debt are placing enormous pressure on government to cut spending and make every dollar count. Every corner of government needs to do its part to spend with great care. With these ingredients in place, we have an unparalleled opening to improve the performance of the federal government. And the opportunity for improvement is significant," he added.

A productivity boom has transformed private-sector performance over the past two decades. As consultants McKinsey & Company and others have pointed out, the Federal Government has almost entirely missed out on this transformation. For example, the Department of Veteran Affairs still largely processes disability claims by hand, passing manila folders 6- to 12-inches thick from metal desktop to metal desktop. Veterans can wait up to 160 days to receive their benefits.

The VA is not alone. The Patent Office, the institution right at the center of protecting and promoting innovation, now receives more than 80 percent of patent applications electronically. That's good. However, these applications are then manually printed out, rescanned and entered into an outdated case management system. The average processing time for a patent is about 3 years. These types of antiquated processes are too common across government. They contribute to the continuing perception that government wastes taxpayer dollars.

Of course, the public sector does face unique challenges including compliance obligations that become real hurdles and objectives beyond the simple bottom-line motivation of the private sector. But many state and local governments, and some federal agencies have been able to work around these constraints and have improved efficiency and raised service quality. The whole federal government has to get on track in order to make that kind of progress but it can—and must—be done.

Kundra has outlined six performance strategies: eliminate waste, drive top priorities, leverage purchasing scale, close the IT performance gap, open government to get results, and attract and motivate top talent. According to Kundra, these are the six strategies that represent the biggest opportunity to boost performance and get government working for the American people.

Remember

Starting with strategy one, eliminate waste. The most sustainable way to save is not to trim around the margins but to cut what doesn't work, what is duplicative, and what is outdated. Through the line-by-line review of the 2010 budget, 121 programs were identified for termination or reduction with savings totaling $17 billion. The Washington Times congratulated the president for the administration's success in discretionary budget cuts in 2010, noting that it was higher than any reductions under the prior administration.

For the 2011 budget, the president proposed 126 additional program cuts totaling $23 billion. To make good choices about where to invest and where to cut going forward, we need a systemic way to evaluate what works and what doesn't. To this end, the president placed a major emphasis on increased funding for rigorous program evaluations in his 2011 budget. These evaluations will help agencies find out whether they're getting the most bang for their buck.

If there are 40 different job training programs going across seven different agencies, where are we getting the greatest impact? Programs that are effective should continue, and those that aren't should either be fixed or terminated. However, wasteful spending isn't just about ineffective programs. In 2009, the federal government reported improper payments of $100 billion. These were payments to the wrong person, the wrong entity, or for the wrong amount. $100 billion of waste is not just a waste of money, it also erodes citizen trust. It is unacceptable.

Remember

The administration has committed to reducing this waste quickly and has set hard targets. Agencies are designating senior accountable officials to own specific reduction targets. It is partnering with states by investing in state pilots to reduce error rates, and is setting up incentives for states to bring their rates down. Agencies are moving aggressively. For example, the Department of Education is reducing Pell Grant payment errors by half of a billion dollars by transferring data directly from the IRS to the students' applications.

Across all these efforts, the administration is actively engaging its partners in Congress. Doing so ensures that the drive to eliminate waste is not viewed solely as an executive branch initiative but rather as a common goal across government. Beyond cutting where we are not getting our money's worth, we also need to make sure that the government is doing what we want it to do, said Kundra.

The administration's next strategy is to drive its top priorities. In most organization, leaders set priorities and then drive the organization to meet these goals. This is hard in government because senior political leaders tend to focus on policy development in crisis management – not execution and implementation. To focus senior agency leaders on getting the most important things done, the High Priority Performance Goal initiative was launched in June 2009.

Agency heads committed to a limited number of goals with high value to the public. These are front and center in the president's FY 2011 budget. Agency leadership, secretaries, and deputy secretaries have taken real ownership here. These management goals set ambitious targets to be achieved within 24 months. The targets are quantifiable. They are well defined and they are outcomes-based.

In looking at the High Priority Performance Goals, there are three attributes. First, the goals are aligned closely with the agencies' missions. For example, the Department of Labor has a goal to train more than 120,000 Americans for green jobs by June 2012. Second, many goals span across agencies, attacking the problem that the government too often works in tight silos, in programs, bureaus, and departments. HUD and VA are not. They're working together with a shared goal to reduce the homeless veteran population by 70 percent by the end of 2012.

Third, and finally, there are many goals to improve the quality of customer and citizen-facing services. All of the available data suggests that the best way to change someone's impression of government is through favorable, direct interactions. To that end, the United States Citizen and Immigration Service has a high priority goal of moving 40 percent of its service delivery online by 2011. This will bring transparency and speed to a process that has historically frustrated applicants.

These High Priority Performance Goals are reviewed regularly with immediate course correction if things are off track. The first two strategies are about concentrating on what matters – cutting what's not working and then focusing on a few key priorities. To support these initiatives, a strong infrastructure is needed – most notably in contracting and information technology. Currently, they both have big problems. As soon as they are fixed, there will be immediate savings and a strong platform to get other things done.

With contracting, government needs to leverage its purchasing scale. The federal government is far and away the world's largest purchaser. It buys over $500 billion of goods and services every year. Despite this scale, too often it does not get the best prices or value for money. Additionally, the federal contracting processes are slow and cumbersome. Even though contracting in the federal government differs from the private sector, it doesn't make it any less important to figure out what is working, what is not working, and reform it.

The president has committed to saving $40 billion through contracting reform by the end of FY 2011. This serves as an important catalyst for action. Government needs to leverage its purchasing power and buy smarter. It needs to work across agencies to take advantage of scale. Take the simple example of office supplies. Over 100 federal organizations have separate contracts for office supplies. As a result, they're paying 30 to 50 percent different prices on any given day for the exact same pens and paper.

The federal government does its purchasing like it is 100 medium-sized businesses, not the world's single largest purchaser. By standardizing specs for commonly purchased items and working together across silos, it can pool its purchasing power to leverage its enormous size and lower cost. But it's not just about pooling purchasing power. It also needs to build up the capacity and the capability of the contracting work force. While dollars spent on contracting doubled during the Bush administration, the size of the work force responsible for managing federal contracts remained flat.

Remember

The growth in contracting volume has outstripped the capacity and capabilities of these professionals. The contracting work force lacks the bandwidth to coordinate effectively with program managers. As a result, contracts are often awarded before the government has figured out exactly what it needs and without full competition. This leads to cost overruns, delays, and dissatisfaction about performance. The FY 2011 budget requested $158 million for agencies to build the capacity and capability of the contracting work force.

Must Do

Given the hundreds of billions of dollars in play here each and every year, this small investment can have a very high return. This investment can save the government tens of billions of dollars and create better results. The president is committed to saving $40 billion from contracting by the end of FY 2011. The administration identified $19 billion in savings for 2010. It is important to fix contracting because it saves money, and because so many other things are dependent on getting contracting right.

Remember

The same is true for information technology. "The gap between where we are and where we need to go is significant," said Kundra. "In fact, I believe IT represents the largest gap between the private and public sectors. Technology has been at the center of those private-sector productivity gains across the past two decades – both efficiency gains and service quality improvements. For the most part, the federal government hasn't participated in these gains."

"In service quality, we're falling further and further behind. If you can book dinner for an airline flight online, then why shouldn't you be able to make an appointment at the local social security office the same way? On the efficiency front, the story is the same. We have antiquated systems and processes throughout many agencies.

"For example, the government system for managing retirement records is stuck in a different era. Here it is, a cave in Boyers, Pennsylvania. Yes, a cave and, yes, the retirement records are stored in 35,000 metal file cabinets. It reminds me of that famous last scene at the end of *The Raiders of the Lost Ark*," said Kundra.

No No

"Clearly there is a better answer here. However, it's not as if we're not trying. There've been several attempts to move the retirement records from the cave era to the digital era. And each time, the efforts have come up short. Unfortunately, this is not an isolated example. Across government we spent $76 billion in information technology last year. And we've spent more than $500 billion across the past decade. But we've got a poor track record of getting our money's worth. When we try to upgrade systems, projects frequently run behind schedule or over budget.

"Too often, projects never accomplish what they were intended to do. To close the IT gap, we need to change how we manage and spend IT dollars. To that end, the president appointed the first ever Federal CIO and Federal CTO to lead and coordinate our technology investments across agencies. We've also launched the IT Dashboard, where leaders in government and the American people can monitor every technology dollar we spend on major projects," Kundra added.

Every major IT project is rated against performance expectations and the administration has launched technical status review sessions and accountability sessions. If a project is over budget, behind schedule, or not performing up to expectations, it will either develop a credible turnaround plan or will be terminated.

At the same time, the administration is also focused on the types of solutions it is pursuing – looking to take advantage of more agile, light technology such as cloud computing, and to shared services wherever possible. If chosen well, these new technologies can be cheaper, faster, and less risky than the big, custom solutions that have too often failed in the past.

Because we all now operate in an increasingly interconnected and complex environment, the administration is also pursuing a comprehensive cyber security approach to securing its digital infrastructure. This means integrating cyber security into the designs of systems rather than bolting it on as an afterthought.

Must Do

Open Government

The president has committed to an unprecedented level of openness. In terms of performance, opening government achieves two things. First, it makes government more accountable and second, it accelerates innovation by engaging the best minds to get to the best solutions.

A good example of accountability is usaspending.gov. This web site aggregates every dollar the federal government spends, whether it's a contract, a grant or other form of assistance. For example, a user is able to compare data on a state-by-state basis and also drill down to the congressional district. Data is sortable by the type of contract award, by size, and by recipient. As new features are added, the administration has focused on improving data quality. It is working with agencies to integrate rigorous data quality checks into the regular financial reviews. This kind of openness leads to accountability for the administration, for federal managers, and for Congress which results in a better use of taxpayer dollars.

Remember

However, to get the most out of people, government has to do more to attract and motivate top talent. The federal government's human resources practices are based on a personnel system that was created 60 years ago. Many of its practices are bureaucratic, cumbersome, and outdated.

Too often, it doesn't focus on people as a primary tool for achieving its missions, and it underinvests in training and development. To attract and retain the best people, there needs to be a fundamental rethink about how employees are both hired and developed.

The hiring process at HUD is a 40-step process. Nineteen different signatures are required, and it takes 139 days from start to finish. Not surprisingly, this results in terrible satisfaction scores from both managers and applicants. And HUD's 139 days-to-hire is not the exception, it's right at the average across all agencies.

The best people don't loiter for five months. They find another home. The good news is that HUD is taking aggressive actions to fix this problem. The administration is also supporting OPM as it works with other agencies' leadership to streamline the hiring process across government.

The goal is to cut the hiring time at least in half by focusing on making the process more candidate-friendly and less bureaucratic. Starting with short, plain language job descriptions, not 20-page documents full of government lingo. Requesting resumes and cover letters, not burdensome essays that don't reliably predict performance.

Remember

Creating a transparent process makes application status clear, not a black hole process that turns off applicants. Most importantly, hiring managers, not just human resources departments, will be held accountable for successfully recruiting the right people for the job.

The next problem is what happens when they get arrive? The federal government has real problems engaging and retaining the best talent. Out of 1.9 million civil servants, only 12,000 are rated below fully successful. That's less than 1 percent. Not surprisingly, only 29 percent of employees believe that adquate steps are taken to deal with poor performers.

Beyond appraisals, there needs to be investment in development. Training and development for frontline workers is vital to achieving efficiency and service quality gains. Given the difficult budget environment, the effectiveness of every dollar spent on training has to be maximized. New hiring systems are needed that attract the best talent with performance systems that motivate, and development and training systems that strengthen them.

The Future is Cloud Computing

A solid majority of technology experts and stakeholders participating in the fourth Future of the Internet survey expect that by 2020 most people will access software applications online and share and access information through the use of remote server networks, rather than depending primarily on tools and information housed on their individual, personal computers. They say that cloud computing will become more dominant than the desktop in the next decade. In other words, most users will perform most computing and communicating activities through connections to servers operated by outside firms.

Among the most popular cloud services now are social networking sites (the 500 million people using Facebook are being social in the cloud), webmail services like Hotmail and Yahoo mail, microblogging and blogging services such as Twitter and WordPress, video-sharing sites like YouTube, picture-sharing sites such as Flickr, document and applications sites like Google Docs, social-bookmarking sites like Delicious, business sites like eBay, and ranking, rating and commenting sites such as Yelp and TripAdvisor.

This does not mean, however, that most of these experts think the desktop computer will disappear soon. The majority sees a hybrid life in the next decade, as some computing functions move towards the cloud and others remain based on personal computers.

Remember

A highly engaged, diverse set of respondents to an online, opt-in survey included 895 technology stakeholders and critics. The study was fielded by the Pew Research Center's Internet & American Life Project and Elon University's Imagining the Internet Center. Some 71% agreed with this statement:

- "By 2020, most people won't do their work with software running on a general-purpose PC. Instead, they will work in internet-based applications such as Google Docs, and in applications run from smartphones. Aspiring application developers will develop for smartphone vendors and companies that provide internet-based applications, because most innovative work will be done in that domain, instead of designing applications that run on a PC operating system."

Some 27% agreed with the opposite statement, which posited:

- "By 2020, most people will still do their work with software running on a general-purpose PC. internet-based applications like Google Docs and applications run from smartphones will have some functionality, but the most innovative and important applications will run on (and spring from) a PC operating system. Aspiring application designers will write mostly for PCs."

Most of those surveyed noted that cloud computing will continue to expand and come to dominate information transactions because it offers many advantages, allowing users to have easy, instant, and individualized access to tools and information they need wherever they are, locatable from any networked device. Some experts noted that people in technology-rich environments will have access to sophisticated, yet affordable, local networks that allow them to "have the cloud in their homes."

Remember

Most of the experts noted that people want to be able to use many different devices to access data and applications, and – in addition to the many mentions of smartphones driving the move to the cloud – some referred to a future featuring many more different types of networked appliances. A few mentioned the "internet of things" – or a world in which everyday objects have their own IP addresses and can be tied together in the same way that people are now tied together by the internet. So, for instance, if you misplace your TV remote, you can find it because it is tagged and locatable through the internet.

Some experts in this survey said that, for many individuals, the switch to mostly cloud-based work has already occurred, especially through the use of browsers and social networking applications. They point out that many people today are primarily using smartphones, laptops, and

desktop computers to network with remote servers and carry out tasks such as working in Google Docs, following web-based RSS (really simple syndication) feeds, uploading photos to Flickr and videos to YouTube, doing remote banking, buying, selling, and rating items at Amazon.com, visiting with friends on Facebook, updating their Twitter accounts, and blogging on WordPress.

Many of the people who agreed with the statement that cloud computing will expand as the internet evolves said the desktop will not die out—it will be used in new, improved ways in tandem with remote computing. Some survey participants said they expect that a more sophisticated desktop-cloud hybrid will be people's primary interface with information. They predicted the desktop and individual, private networks will be able to provide most of the same conveniences as the cloud but with better functionality, overall efficiency, and speed. Some noted that general-purpose in-home PC servers can do much of the work locally via a connection to the cloud to tap into resources for computing-intensive tasks.

Among the defenses for a continuing domination of the desktop, many said that small, portable devices have limited appeal as a user interface and they are less than ideal for doing work. They also expressed concern about the security of information stored in the "cloud" (on other institutions' servers), the willingness of cloud operators to handle personal information in a trustworthy way, and other problems related to control over data when it is stored in the cloud, rather than on personally-controlled devices.

Some respondents observed that putting all, or most of, faith in remotely accessible tools and data puts a lot of trust in the humans and devices controlling the clouds and exercising gatekeeping functions over access to that data. They expressed concerns that cloud dominance by a small number of large firms may constrict the internet's openness and its capacity to inspire innovation – that people are giving up some degree of choice and control in exchange for streamlined simplicity. A number of people said cloud computing presents difficult security problems and further exposes private information to governments, corporations, thieves, opportunists, and human and machine error.

Survey participants noted that there are also service quality and compatibility hurdles that must be crossed successfully before cloud computing gains more adopters. Among the other limiting factors the expert respondents mentioned were: the lack of broadband spectrum to handle the load if everyone is using the cloud; the variability of cost and access in different parts of the world, and the difficulties that lie ahead before they can reach the ideal of affordable access anywhere, anytime; and complex legal issues, including cross-border intellectual property and privacy conflicts.

Among the other observations made by those taking the survey were: large businesses are far less likely to put most of their work "in the cloud" anytime soon because of control and security issues; most people are not able to discern the difference between accessing data and applications on their desktop and in the cloud; low-income people in least-developed areas of the world are most likely to use the cloud, accessing it through connection by phone.

Source: "The Future of Cloud Computing," Pew Research Center's Internet & American Life Project and Elon University's Imagining the Internet Center (June 2010)

Step One. What is Cloud Computing?

Definition: Cloud computing is a model for enabling ubiquitous, convenient, on-demand network access to a shared pool of configurable computing resources (e.g., networks, servers, storage, applications, and services) that can be rapidly provisioned and released with minimal management effort or service provider interaction.

Source: National Institute of Standards and Technology (NIST).

Five Key Characteristics

Rapid Elasticity

Elasticity is defined as the ability to scale resources both up and down as needed. To the consumer, the cloud appears to be infinite, and the consumer can purchase as much or as little computing power as they need. This is one of the essential characteristics of cloud computing in the NIST definition.

Measured Service

In a measured service, aspects of the cloud service are controlled and monitored by the cloud provider. This is crucial for billing, access control, resource optimization, capacity planning, and other tasks. Cloud systems automatically control and optimize resource use by leveraging a metering capability at some level of abstraction appropriate to the type of service (e.g., storage, processing, bandwidth, and active user accounts). Resource usage can be monitored, controlled, and reported providing transparency for both the provider and consumer of the utilized service.

On-Demand Self-Service

The on-demand and self-service aspects of cloud computing mean that a consumer can use cloud services as needed without any human interaction with the cloud provider.

Ubiquitous Network Access (Broad Network Access)

Ubiquitous network access means that the cloud provider's capabilities are available over the network and can be accessed through standard mechanisms by both thick and thin clients (e.g., mobile phones, laptops, and PDAs).

Resource Pooling

Remember

Resource pooling allows a cloud provider to serve its consumers via a multi-tenant model. Physical and virtual resources are assigned and reassigned according to consumer demand. There is a sense of location independence in that the customer generally has no control or knowledge over the exact location of the provided resources but may be able to specify location at a higher level of abstraction (e.g., country, state, or datacenter). Examples of resources include storage, processing, memory, network bandwidth, and virtual machines.

Rapid Elasticity

Capabilities can be rapidly and elastically provisioned, in some cases automatically, to quickly scale up and rapidly released to quickly scale down. To the consumer, the capabilities available for provisioning often appear to be unlimited and can be purchased in any quantity at any time.

Four Deployment Models

Public Cloud

In simple terms, public cloud services are characterized as being available to clients from a third-party service provider via the internet. The term "public" does not always mean free, even though it can be free or fairly inexpensive to use. A public cloud does not mean that a user's data is publically visible—public cloud vendors typically provide an access control mechanism for their users. Public clouds provide an elastic, cost-effective means to deploy solutions.

Although many organizations use public clouds for private business benefit, they don't control how those cloud services are operated, accessed, or secured. Popular examples of public clouds include Amazon EC2, Google Apps, and Salesforce.com.

Many organizations have adopted different cloud models simultaneously, leading to a hybrid cloud environment in which some IT assets and services are hosted in internal clouds while others are delivered through externally hosted private clouds and public clouds.

Benefits of Public Clouds

Public Cloud benefits include:

- ☐ Reduced computing costs
- ☐ Reduced infrastructure footprint
- ☐ Achievement of a more flexible computing environment
- ☐ Map capacity to demand (no under- or over-provisioning – buy only what is needed)
- ☐ Decoupling of applications from infrastructure constraints
- ☐ Ensured capacity is there when you need it

Private Cloud

Private cloud describes an IT infrastructure in which a shared pool of computing resources—servers, networks, storage, applications, and software services—can be rapidly provisioned, dynamically allocated and operated for the benefit of a single organization. A private cloud offers many of the benefits of a public cloud computing environment, such as being elastic and service-based. The difference between a private cloud and a public cloud is that in a private cloud-based service, data and processes are managed within the organization without the restrictions of network bandwidth, security exposures, and legal requirements that using public cloud services might entail. In addition, private cloud services offer the provider and the user greater control of the cloud infrastructure, improving security and resiliency because user access and the networks used are restricted and designated.

Private cloud describes an IT infrastructure in which a shared pool of computing resources—servers, networks, storage, applications, and software services—can be rapidly provisioned, dynamically allocated, and operated for the benefit of a single organization. Private clouds are similar in many ways to the traditional IT service delivery model, with three key differences:

- ☐ IT resources are virtualized, leading to much more efficient use and flexible allocation. Most notably, virtualization enables the dynamic transfer or sharing of services within the cloud infrastructure, and the secure partitioning of services for multitenancy. "Tenants" sharing a server or application can be either completely different companies in an external cloud scenario or different business functions or groups within internal clouds.
- ☐ The organization does not need to physically own or operate the IT assets that form its private cloud. Some assets can be outsourced to cloud providers. For instance, outside data centers may be leased to run specific applications. Nevertheless, the organization still effectively "owns" its private cloud by controlling and setting policies governing how virtual IT assets are operated, with cloud vendors guaranteeing specific levels of service, and conformance to agreed-upon standards for information security and compliance.
- ☐ Within a virtualized environment, just about everything can be measured, including CPU cycles and bits transmitted. As a result, clouds can be monitored at a highly granular level beyond the typical latency- and performance-based measurements of traditional IT environments. This opens up the potential for usage-based billing or charge backs already common with public cloud services, such as Amazon's Elastic Compute Cloud (EC2).

Internal clouds are a type of private cloud in which all aspects of IT service delivery are physically owned and operated by the organization itself. In terms of monitoring and proving compliance with information policies, organizations presumably have complete visibility, transparency, and control over their internal clouds because they own and maintain the entire cloud infrastructure, from servers to services.

Benefits of Private Clouds

Benefits of private clouds include:

- ☐ Reduced TCO for operating infrastructure – hardware, power, cooling
- ☐ Increased ROI on existing hardware
- ☐ Achievement of a more flexible computing environment
- ☐ Map capacity to demand (no under- or over-provisioning – buy only what's needed)
- ☐ Automation of manual provisioning tasks
- ☐ Decoupling of applications from infrastructure constraints
- ☐ Ensured capacity is there when you need it

Hybrid Cloud

A hybrid cloud is a combination of a public and private cloud that interoperates. In this model, users typically outsource non-business-critical information and processing to the public cloud, while keeping business-critical services and data in their control.

Benefits of Hybrid Clouds

Benefits of Hybrid clouds include:

- ☐ Achievement of a more flexible computing environment
- ☐ Reduced computing costs
- ☐ Map capacity to demand (no under- or over-provisioning – buy only what's needed)
- ☐ Automated manual provisioning tasks
- ☐ Decoupling of applications from infrastructure constraints
- ☐ Ensured capacity is there when you need it

Figure 1. Cloud Sourcing Models

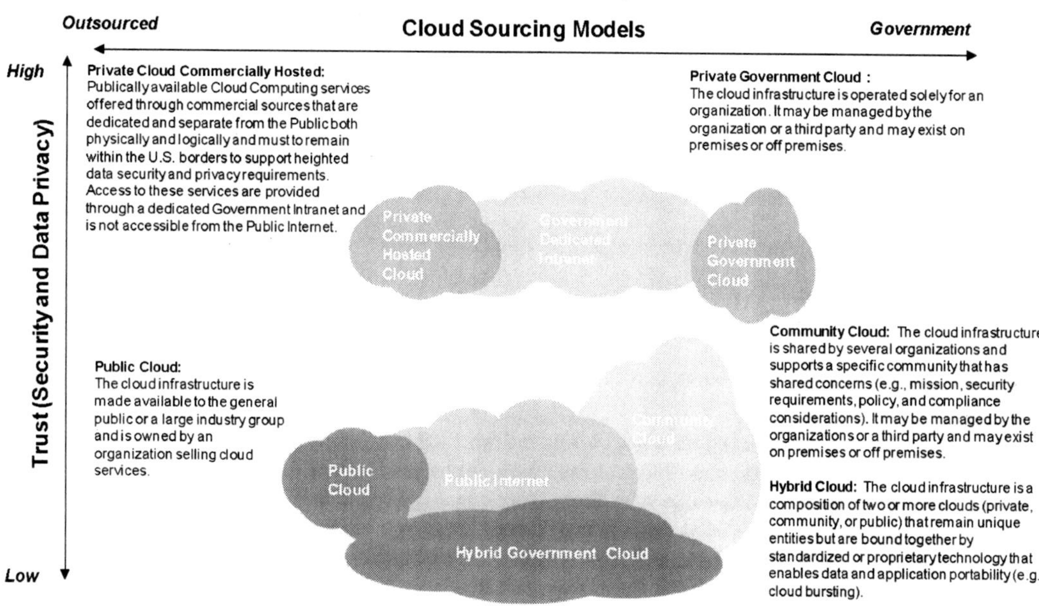

Community Cloud

A community cloud is controlled and used by a group of organizations that have shared interests, such as specific security requirements or a common mission. The members of the community share access to the data and applications in the cloud.

Cloud Adoption Trends

A little known fact is that most government cloud projects actually leverage public cloud services. Since over 90% of all data owned by the Federal government is public, implementation of many constituent support functions was relatively easy and involved very low risk. Private clouds, however, offer the greatest near-term value in increasing efficiency and lowering operational cost. These implementations deliver immediate benefits while still preserving control over the applications and information residing in their private cloud. Some organizations are already looking to the next step: integrating external technology providers to further enhance the service capabilities and operational efficiency of their cloud environments. As the prospective benefits of leveraging external service providers continue to grow, many enterprise clouds will integrate outside cloud infrastructure or platform services. Some will even integrate whole public cloud services to create new hybrid models of cloud computing.

The rise of external service providers introduces new complexities, as well as benefits, into the delivery chain for cloud services. The leading areas of concern, according to a survey from IDG Research Services, relate to managing and safeguarding corporate information in clouds with externally hosted components. This is particularly true as information and application control moves off-premise to third-party providers. Without early planning and consideration, the evolution of more complex hybrid models could lead to the following concerning conditions, which are weighing on the minds of today's CIOs and IT professionals:

- ☐ Growth and proliferation of incompatible cloud services
- ☐ Isolation of valuable corporate information within cloud-based silos
- ☐ Escalating potential for vendor lock-in
- ☐ New complications in enforcing information security and policy compliance

Must Do

If not planned for, these emerging conditions will impede the flow and value of corporate information for years to come. The main driver behind these problems is the classic challenge of information silos – the lack of cloud interoperability standards, lack of shared services that underpin multiple applications, and lack of tools to access information across applications. Like in on-premise installations, these gaps present a serious challenge in sharing information between applications, systems, and even across cloud environments. Today's solutions providers are busy tackling the barriers to interoperability: embracing open standards, building shared services, and creating standardized technology platforms, as well as creating APIs that help automate the integration of services across clouds.

Although solution providers are making progress on the technology front, CIOs still see a gap between where cloud services are and where they need to be, particularly when they involve outside service providers. While they look forward to a time when standards have evolved and cloud platforms are fully enterprise ready, they don't want to sit on the sidelines waiting for cloud maturity and lose the competitive and cost advantages of moving IT services to the cloud today. Additionally, business units in some organizations are forcing IT's hand by independently provisioning public cloud services – sealing the deal with a corporate credit card and a user terms and conditions checkbox. As a result, organizations increasingly find themselves supporting hybrid environments in which some information resides in public clouds, some resides in externally hosted private clouds and some resides within the enterprise, either in traditional or virtualized data centers.

The net result is that organizations are at risk of fragmenting their information architecture, isolating valuable corporate information within disparate applications and cloud services. It is not unlike the enterprise data silo problems of the '90s, when information was locked within ERP and CRM applications, requiring massive systems integration efforts. The business consequence of siloed data is more urgent in today's climate: if you can't access and use your information, you've surrendered your business agility and lost an important competitive edge.

Source: Creating Information Advantage in a Cloudy World, Leadership Council for Information Advantage.

Three Delivery Models

Software-as-a-Service (SaaS)

The consumer uses an application, but does not control the operating system, hardware or network infrastructure on which it is running. Example: Salesforce.com

Platform-as-a-Service

The consumer uses a hosting environment for their applications. The consumer controls the applications that run in the environment (and possibly has some control over the hosting environment), but does not control the operating system, hardware or network infrastructure on which they are running. The platform is typically an application framework. Examples: Google AppEngine, Force.com

Infrastructure as a Service (IaaS)

The consumer uses fundamental computing resources such as processing power, storage, and networking components or middleware. The consumer can control the operating system, storage, deployed applications, and possibly networking components such as firewalls and load balancers, but not the cloud infrastructure beneath them. Example: Amazon played a key role in the development of cloud computing by modernizing their data centers after the dot-com bubble, which, like most computer networks, were using as little as 10% of their capacity at any one time just to leave room for occasional spikes. Having found that the new cloud architecture resulted in significant internal efficiency improvements whereby small, fast-moving, "two-pizza teams" could add new features faster and easier, Amazon started providing access to their systems through Amazon Web Services on a utility computing basis in 2005. The genesis of Amazon Web Services has been characterized as an extreme over-simplification by a technical contributor to the Amazon Web Services project.

Cloud Computing Is Not:

- ☐ Grid Computing - a form of distributed computing, whereby a 'super and virtual computer' is composed of a cluster of networked, loosely coupled computers acting in concert to perform very large tasks
- ☐ Utility Computing - the packaging of computing resources, such as computation and storage, as a metered service similar to a traditional public utility, such as electricity

General Cloud Computing Benefits

There was a time when every household, town, farm or village had its own water well. Today, shared public utilities give us access to clean water by simply turning on the tap—cloud computing

works in a similar fashion. Just like the water from the tap in your kitchen, cloud computing services can be turned on or off quickly, as needed. Like the water company, there is a team of dedicated professionals making sure the service provided is safe and available on a 24/7 basis. Best of all, when the tap is not on, not only are you saving water, but you aren't paying for resources you do not currently need.

Economical

Remember

Cloud computing is available at a fraction of the cost of traditional IT services, eliminating upfront capital expenditures and dramatically reducing administrative burden on IT resources. Cloud computing is a pay-as-you-go approach to IT, in which only a low initial investment is required to get going. Additional investment is incurred as system use increases, and costs can decrease if usage decreases. Because of this, cash flows better match total system cost.

Flexible

IT departments that anticipate fluctuations in user load do not have to scramble to secure additional hardware and software. With cloud computing, they can add and subtract capacity as network load dictates, and pay only for what they use. Cloud computing provides on-demand computing across technologies, business solutions, and large ecosystems of providers, reducing time to implement new solutions from months to days.

Rapid Implementation

Without the need to go through the procurement and certification processes, and with a near-limitless selection of services, tools, and features, cloud computing helps projects get off the ground in record time.

Consistent Service

Network outages can send an IT department scrambling for answers. Cloud computing can offer a higher level of service and reliability, and an immediate response to emergency situations.

Increased Effectiveness

Cloud computing frees the user from the finer details of IT system configuration and maintenance, enabling them to spend more time on mission-critical tasks, and less time on IT operations and maintenance.

Energy Efficient

Because resources are pooled, each user community does not need to have its own dedicated IT infrastructure. Several groups can share computing resources, leading to higher utilization rates, fewer servers, and less energy consumption.

Access anywhere

You are no longer tethered to a single computer or network. You can change computers or move to portable devices, and your existing applications and documents follow you through the cloud.

Elastic Scalability and Pay-as-You-Go

Add and subtract capacity as your needs change. Pay for only what you use.

Easy to Implement

You do not need to purchase hardware, software licenses, or implementation services.

Remember

Service Quality

Cloud service providers offer reliable services, large storage and computing capacity, and 24/7 service and up-time.

Delegate Non-Critical Applications

Cloud computing provides a way to outsource non-critical applications to service providers, allowing agency IT resources to focus on business-critical applications.

Always the Latest Software

You are no longer faced with choosing between obsolete software and high upgrade costs. When the applications are web-based, updates are automatic and are available the next time you log into the cloud.

Tips

Sharing Documents and Group Collaboration

Cloud computing lets you access all your applications and documents from anywhere in the world, freeing you from the confines of the desktop and facilitating group collaboration on documents and projects.

Advantages of Cloud Computing for the Federal Government

Cloud computing offers a cost-effective, service-oriented approach for sharing computing resources, whereby common infrastructure, applications, information, and solutions can be utilized across the Government. The overall objective is to create a more agile Federal enterprise, where services can be provisioned and reused on demand to meet business needs.

The Federal Cloud Computing Initiative (FCCI)

The Federal Cloud Computing Initiative (FCCI) is focused on implementing cloud computing solutions for the Federal Government that increase operational efficiencies, optimize common services and solutions across organizational boundaries, and enable transparent, collaborative, and participatory government.

FCCI's basic principles stem from the ITI LoB, a Government-wide study of end-user systems and support, telecommunications, mainframe, and server services including optimizing performance, efficiency, and effectiveness, and year-over-year comparisons. Subsequently, the Federal CIO Council identified cloud computing as a Federal IT priority and formed a CIO Working Group, the Cloud Computing Executive Steering Committee (CCESC), in March of 2009. GSA's Casey Coleman chairs the CCESC, which leads the Federal Cloud Computing Initiative.

The objective of the FCCI is to make cloud computing services accessible and easy to procure for Federal agencies. Actions taken so far have included:

- ☐ Creating a cloud computing definition (NIST)
- ☐ Hosting a cloud computing summit
- ☐ Releasing an IaaS RFI and subsequent IaaS RFQ
- ☐ Launching a cloud computing storefront: Apps.gov

The FCCI Executive Steering Committee, Advisory Council, and Working Groups continue to work on identifying and addressing issues and obstacles to successful cloud computing implementation.

FCCI Mission and Vision

Cloud Computing Mission Statement

Drive the government-wide adoption of cost effective, green, and sustainable Federal cloud computing solutions.

Cloud Computing Vision Statement

Establish secure, easy to use, rapidly provisioned IT services for the Federal Government, including:

- ☐ Agile and simple acquisition and certification processes;
- ☐ Elastic, usage-based delivery of pooled computing resources;
- ☐ Portable, reusable and interoperable business-driven solutions;
- ☐ Browser-based ubiquitous internet access to services; and
- ☐ Always on and available, utility-like solutions.

GSA and FCCI

The General Services Administration (GSA) is participating in the Federal Cloud Computing Initiative and is responsible for the coordination of GSA's activities with respect to the Initiative via its Program Management Office (CC PMO). GSA and the CC PMO are focusing on implementing projects for planning, acquiring, deploying, and utilizing cloud computing solutions for the Federal Government that increase operational efficiencies, optimize common services and solutions across organizational boundaries, and enable transparent, collaborative, and participatory government.

In a major step toward Government-wide adoption of cloud computing services, the U.S. General Services Administration in coordination with the Federal Chief Information Officers Council announced comprehensive requirements for the Federal Risk and Authorization Management Program, or FedRAMP, for public comment. FedRAMP will reduce redundant processes across government by providing security authorizations and continuous monitoring of cloud systems that can be leveraged by agencies to streamline their security process while providing highly effective security services.

"As part of the President's Accountable Government Initiative, we are working to close the IT gap between the private and public sectors, and leverage technology to make government work harder, smarter, and faster for the American people," said Federal CIO, Vivek Kundra. "By simplifying how agencies procure cloud computing solutions, we are paving the way for more cost-effective and energy-efficient service delivery for the public, while reducing the federal government's data center footprint."

FedRAMP was established to provide a standard approach to assessing and authorizing cloud computing services and products. FedRAMP allows joint authorizations and continuous security monitoring services for government and commercial cloud computing systems intended for multiagency use. Joint authorizations of cloud providers will result in a common security risk model that can be leveraged across the federal government, ensuring a consistent baseline for cloud-based technologies.

"Ensuring data and systems security is one of the biggest and most important challenges for federal agencies moving to the cloud," said David McClure, GSA's Associate Administrator for Citizen Services and Innovative Technologies. "FedRAMP's uniform set of security authorizations can eliminate the need for each agency to conduct duplicative, time-consuming, costly security reviews. By going out for public comment, leveraging knowledge from industry, government, and the public, ensures our requirements maximize security while easing access toward the cloud."

Remember

GSA Apps.gov

Cloud computing offerings including Infrastructure as a Service (Iaas), Software as a Service (SaaS), and Platform as a Service (PaaS) are available to customer agencies to acquire from

vendors today. All cloud computing offerings are already available to be procured using full and open competition after development of a Statement of Work or Statement of Objectives. Some cloud computing offerings can be procured today by placing orders against the GSA Schedule 70 in accordance with FAR 8.4, and a subset of these vendors can be found under the categories of Productivity Apps and Business Apps on the Apps.gov website. These vendors do not have BPAs. The offerings identified are available on their current Schedule 70 contracts and have been included on Apps.gov to assist your agency in learning what cloud services are already available. Schedule 70 vendors already selling Software as a Service cloud services are adding themselves to Apps.gov Productivity Apps and Business Apps sections of Apps.gov on an on-going basis. In addition to identifying already awarded Schedule 70 cloud offerings, Apps.gov will soon post offerings from awardees of a Blanket Purchase Agreement for Infrastructure as a Service (IaaS). Awards have not been made but once awards are made, the Cloud IT Services section of Apps.gov will be populated. By awarding cloud computing-specific BPAs, GSA is supporting the President's Cloud Computing Initiative by making it easier for agencies to make the shift to acquiring IT from the cloud. The prices offered on the BPAs will be discounts from vendors' Schedule 70 prices. The Schedule 70 contracts, BPAs and orders placed against each are authorized and regulated by the Federal Acquisition Regulation 48 C.F.R. 1-53, and other customer agency specific acquisition regulations and appropriations laws that may apply.

Social Media Tools:

All of the social media tools the GSA places on their storefront are offered by Web 2.0 providers as free services. Since no payments are involved, and agencies are expending no appropriated funds, government contracting procedures of the FAR do not apply. Instead, the relationship between the social media provider and the user agency will be reflected in a signed Terms of Service (TOS) Agreement.

Social media can take many different forms, including internet forums, social blogs, wikis, podcasts, photos, videos, rating, and bookmarking. Technologies include: blogs, video hosting, photo-sharing, wall-postings, e-mail, instant messaging, and music sharing. Applications are computer programs or services available over the web.

Social Media Providers

A coalition of federal agencies, led by GSA's Office of Citizen Services, began working with a broad range of providers of no-cost social media products to develop amended Terms of Service (TOS) agreements that reflect the needs of federal users.

GSA led this effort because the existing standard Terms of Service on most social media sites do not comply with federal law and, in many ways, are not compatible with agency expectations and practices. Working with providers to amend their standard TOS for federal users eliminates those problems.

Providers who want to offer free social media products under federal-friendly Terms of Service can do so through Apps.gov for the federal government's consideration.

Remember

If an agency chooses to use various social media sites and tools to accomplish its mission, the agency won't have to start from scratch in negotiating with the providers if federal-friendly Terms of Service are offered on Apps.gov. However, each agency must first consider its own policies for the use of social media, its specific needs, expectations, and practices, along with a legal review, before signing the agreement.

While these TOS amendments resolve the major legal issues of the sign-up process, clarify expectations, and set the stage for productive use of these services, agencies must still comply with laws and regulations on security, privacy, accessibility, records retention, ethical use, and other specific agency policies and requirements when they use the tools. This is why we recommend you seek the advice of your agency counsel on whether the TOS is legally appropriate for use by your agency.

Remember

Because these products and services are free and don't involve agency appropriated funds, the agreements are not considered to be government contracts nor are they procurements under the Federal Acquisition Regulation (FAR). In contrast, fee-based products fall under all federal procurement rules and regulations, and are not part of the line-up of social media products on Apps.gov.

Using Social Media Applications

Social media applications are available for consideration by all Federal agencies. Be sure your agency can agree to the legal provisions of each agreement. Because several agencies helped to negotiate these agreements, it is both expected and hoped that most other agencies will find the language acceptable. However, each agency needs to sign its own agreement.

Tips

Think strategically about how these tools can help you accomplish your agency's mission. Social media applications may be used to meet customer needs, gather feedback and new ideas, and disseminate information, images and video to your stakeholders, clients, and citizens. Establishing an agency presence through a social media channel promotes citizen engagement, collaboration and transparency. An example of how the White House reaches out to citizens through a YouTube channel is found here: http://www.youtube.com/whitehouse. GSA has created the U.S. Government YouTube channel at http://www.youtube.com/usgovernment.

At this time, these TOS amendments only cover federal agencies as they were written to comply with federal law. The National Association of State Chief Information Officers (NASCIOs) has told GSA it is working with states to develop a template that appropriately addresses general state requirements, so that NASCIO can begin discussions with individual providers on behalf of states.

The Federal Data Center Consolidation Initiative

The transition to cloud computing is also supported by Federal data center consolidation efforts. The consolidation of Federal data centers will reduce energy consumption, space usage, and environmental impacts, while increasing the utilization and efficiency of IT assets. Data center consolidation will also play an important role in meeting the goals of the Energy Security and Independence Act of 2007 and various executive orders directing increased energy efficiencies. The effort will promote shared Government-wide, cost-effective, green, and sustainable Federal data centers in support of agency missions.

In February 2010, the Federal CIO issued data center consolidation guidance to agencies regarding creation of agency data center consolidation plans. The guidance directed agencies to consider agency data center performance and utilization metrics, energy-efficiency use data, physical facility, operational cost and asset information, best practices, open standards, and security. Agencies will develop their data center consolidation plans and incorporate them into their FY 2012 budgets.

Standards Development

As part of the Federal Cloud Computing Initiative, the National Institute of Standards and Technology (NIST) is leading and facilitating the development of cloud computing standards which respond to high priority security, interoperability, and portability requirements.

NIST Support to Cloud Computing

NIST serves as the government lead, working with other government agencies, industry, academia, Standards Development Organizations (SDO), and others to leverage appropriate existing standards and to develop cloud computing standards where gaps exist. While cloud computing services are currently being used, security, interoperability, and portability are cited as major barriers to further adoption. The expectation is that standards will shorten the adoption cycle, enabling cost savings and an increased ability to quickly create and deploy enterprise applications. The focus is on standards which support interoperability, portability, and security to enable important usage scenarios.

NIST scientific expertise and its diverse group of NIST IT scientists yield a collective knowledge, research, and technical guidance capability which is aligned with the bureau's mission to support industry and advise government, acting impartially, and providing credible technical insights.

In 2009, NIST made the widely adopted and referenced NIST Definition of Cloud Computing publicly available. NIST is in the process of developing a series of Special Publications (SP) related to cloud computing. These Special Publications describe the activities which are listed next.

As part of its Technical Advisory effort NIST will:

- ☐ Provide technical support and leadership to the working groups supporting the Federal CIO Council
- ☐ Create guidance to facilitate leveraged Government authorization of cloud systems, and guidance on the application of FISMA and 800-53 to cloud computing

Standards Acceleration to Jumpstart Adoption of Cloud Computing (SAJACC)

There is often a gap between the time when formal standards for a new technology are needed and when they become available. The development of standards is inherently dependent on the time consuming process of consensus-building through broad participation. There is also a need to ensure due diligence in producing a standard of quality and completeness such that it will be effective and broadly adopted.

The SAJAAC goal is to facilitate the development of cloud computing standards. SAJACC will include a publicly accessible NIST hosted portal which facilitates the exchange of verifiable information regarding the extent to which pre-standard candidate interface specifications satisfy key cloud computing requirements. The expectation is that SAJACC will help to accelerate the development of cloud computing standards and, as a by-product of its information dissemination function, increase the level of confidence to enable cloud computing adoption.

The SAJAAC strategy and approach is to accelerate the development of standards and to increase the level of confidence in cloud computing adoption during the interim period before cloud computing standards are formalized. SAJACC will provide information about interim specifications and the extent that they support key cloud computing requirements through a NIST hosted SAJACC portal. More specifically, SAJACC will provide a public internet-accessible repository of cloud computing usage scenarios (i.e., use cases), documented cloud system interfaces, pointers to cloud system reference implementations, and test results showing the extent to which different interfaces can support individual use cases.

The project is in the process of formulating an initial set of draft use cases and vetting these with cloud computing stakeholders in academia, government, and industry. The use cases are being developed to demonstrate portability, interoperability, and achievable security for users of cloud systems. After the use cases have been refined, they will be published on the portal. The project will then identify candidate legacy cloud system interfaces, along with their reference implementations, for validation against the use cases. After an initial set of legacy interfaces have been identified, NIST will conduct validation tests and publish the results. The process of identifying new interfaces (with corresponding reference implementations) and new use cases will be ongoing.

SAJACC leverages, coordinates, and is heavily dependent on input from all stakeholders with an interest in cloud computing standards.

Federal Risk and Authorization Management Program (FedRAMP):

NIST's role is to support the definition of a consistent technical process that will be used by FedRAMP to assess the security posture of specific cloud service implementations. NIST serves as a technical advisor for the FedRAMP process that will be implemented by the Federal CIO Council.

NIST, in the technical advisory role to the interagency Federal Cloud Computing Advisory Council (CCAC) Security Working Group will define an initial technical approach and process for FedRAMP consistent with NIST security guidance in the context of the Federal Information System Management Act (FISMA). To clarify the role of NIST with respect to FedRAMP: while NIST is supporting the definition of the FedRAMP process from a technical perspective, NIST is not the implementing organization. The governance and operational implementation of FedRAMP will be completed under the auspices of the Federal CIO Council.

Special Publications on Cloud Computing and Selected Topics

Remember

NIST plans to issue an initial SP on cloud computing. The purpose is to provide insight into the benefits and considerations, and the secure and effective uses of cloud computing. More specifically, the document will provide guidance on key considerations of cloud computing: interoperability, portability, and security. To present these issues, the document will use the broadly recognized and adopted NIST Definition of Cloud Computing as a basis, given informal models of the major cloud computing service categories (Software as a Service, Platform as a Service, and Infrastructure as a Service). The publication will outline typical terms of use for cloud systems, will synopsize future research areas in cloud computing, and will provide informal recommendations.

NIST is also in the process of developing an SP on securing virtualization solutions for servers and desktops which are widely used in cloud computing technologies. The publication will provide an overview of full virtualization technologies, discuss the security concerns associated with full virtualization for servers, and provide recommendations for addressing them. The publication will also give an overview of actions that organizations should perform throughout the lifecycle of a server virtualization solution.

U.S. Government Use Cases

Government Cloud Computing Initiatives

Cloud computing pilots are being initiated by many U.S. government agencies. Governments from around the world are also starting to take notice of the flexibility, operational benefits, and substantial cost savings that cloud computing can provide. This section highlights some of the government programs that are utilizing cloud computing.

United States

- ☐ Federal Chief Information Officers Council
  - Data.gov and IT Dashboard
    - Cloud computing plays a key role in the President's initiative to modernize Information Technology (IT) by identifying enterprise-wide common services and solutions, and adopting a new cloud computing business model. The Federal CIO Council, under the guidance of the Office of Management and Budget (OMB), and the Federal Chief Information Officer (CIO), Vivek Kundra, established the Cloud Computing Initiative to fulfill the President's objectives for cloud computing.
- ☐ Defense Information Systems Agency (DISA)
  - The DISA is developing a number of cloud computing solutions available to U.S. military, DoD government civilians, and DoD contractors for Government authorized use. They include: Forge.mil, a system that currently enables the collaborative development and use of open source and DoD community source software; GCDS, a commercially-owned, globally-distributed computing platform that provides a reliable and secure content and application distribution services solution that delivers applications to dispersed user communities; and RACE, a quick-turn computing solution that uses cloud computing to deliver platforms that are quick, inexpensive, and secure.
- ☐ U.S. Department of Energy
  - Magellan
    - The Magellan program is funded by the American Recovery and Reinvestment Act through the U.S. Department of Energy (DOE), and will examine cloud computing as a cost-effective and energy-efficient computing paradigm for scientists to accelerate discoveries in a variety of disciplines. To test cloud computing for scientific capability, DOE centers at the Argonne Leadership Computing Facility (ALCF) in Illinois and the National Energy Research Scientific Computing Center (NERSC) in California will install similar mid-range computing hardware, but will offer different computing environments. The combined set of systems will create a cloud testbed that scientists can use for their computations while also testing the effectiveness of cloud computing for their particular research problems.
- ☐ General Services Administration (GSA)
  - Apps.gov
    - The GSA is focusing on implementing projects for planning, acquiring, deploying, and utilizing cloud computing solutions for the Federal Government that increase operational efficiencies, optimize common services and solutions across organizational boundaries, and enable transparent, collaborative, and participatory government.

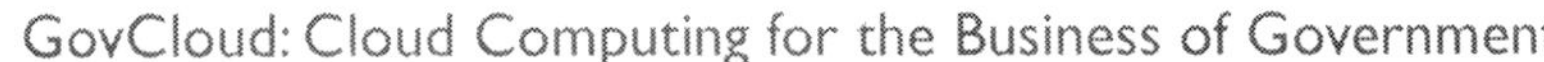

This includes Apps.gov, the official Cloud Computing Storefront for the Federal Government. The site features a complete listing of all GSA-approved cloud services available to federal agencies.

- ☐ Department of the Interior
  - National Business Center (NBC) Cloud Computing
    - The Department of the Interior's National Business Center (NBC) plans to bring the benefits of cloud computing to both NBC's business services clients and data center hosting clients alike through advancements to the highly efficient NBC shared infrastructure. The NBC is offering six cloud computing products: NBCGrid (IaaS), NBCFiles (Cloud Storage), NBCStage (PaaS), NBC Hybrid Cloud, NBCApps (SaaS Marketplace), and NBCAuth.
- ☐ NASA
  - Nebula
    - Nebula is a cloud computing pilot under development at NASA Ames Research Center. It integrates a set of open-source components into a seamless, self-service platform, providing high-capacity computing, storage, and network connectivity using a virtualized, scalable approach to achieve cost and energy efficiencies.
- ☐ National Institute of Standards and Technology (NIST)
  - NIST's role in cloud computing is to promote the effective and secure use of the technology within government and industry by providing technical guidance and promoting standards.

United Kingdom

- ☐ G-Cloud
  - The U.K. Government's CIO announced the establishment of a UK onshore, private Government Cloud Computing Infrastructure called G-Cloud. In essence, the program will include Infrastructure as a Service (IaaS), Middleware/Platform as a Service (PaaS) and Software as a Service (SaaS). In relation to SaaS, the government would establish a Government Application Store. This plan is supported by the U.K. Government's Digital Britain initiative.

European Union

- ☐ Resources and Services Virtualization without Barriers Project (RESERVOIR)
- ☐ EUROCLOUD
- ☐ Seventh Framework Programme

The Seventh Framework Programme (FP7) bundles all research-related EU initiatives together under a common roof, playing a crucial role in reaching the goals of growth, competitiveness, and employment. The FP7 is funding several projects on cloud computing and has also compiled a group of experts to outline the future direction of cloud computing research.

Canada

- ☐ Canada Cloud Computing
  - ■ The Canadian Government's CTO of Public Works Government Services presented a paper on cloud computing and the Canadian Environment. This paper essentially outlines the Canadian Government's considerations of cloud computing by outlining the advantage of their cold landscape (among other things) as a prime location for the construction of large energy efficient cloud computing data centers.

Japan

- ☐ The Digital Japan Creation Project (ICT Hatoyama Plan)
- ☐ The Kasumigaseki Cloud
  - ■ Japan's Ministry of Internal Affairs and Communications (MIC) released a report outlining the Digital Japan Creation Project (ICT Hatoyama Plan) which seeks to create new Information and Communications Technology (ICT) markets to help boost Japan's economy. Within this plan is an outline to create a nationwide cloud computing infrastructure tentatively called the Kasumigaseki Cloud.

More detailed examples of cloud computing in action at U.S. federal and state levels are given at the end of Step Four.

NIST will lead interested USG agencies and industry to define target USG Cloud Computing business use cases (set of candidate deployments to be used as examples) for cloud computing model options to identify specific risks, concerns, and constraints. For example, a candidate deployment might be employee e-mail, or migration of a specific application system, to a specific cloud computing model option. NIST will lead and facilitate this effort via the Federal CIO Council sponsored Cloud Computing Standards Working Group and working groups by NIST.

NCOIC: Cloud Interoperability

The NCOIC is an international organization for accelerating the global implementation of network-centric principles and systems to improve information sharing among various communities of interest for the betterment of their productivity, interactivity, safety, and security. The NCOIC Cloud Computing Working Group works to collaborate and engage with other cloud groups to look at standards-based solutions with a focus on interoperability. The group also document best practices, architectures, and blueprints for commercially-available implementations

including examining security implications and how to implement an internal cloud. To further the adoption of interoperable cloud technology, a Hybrid Cloud Computing (HCC) pattern is being developed.

NCOIC Hybrid Cloud Computing Pattern

The Hybrid Cloud Computing pattern (HCC) would be used in any environment requiring the use of cloud computing capabilities to support service-oriented Government, and builds on the use and maturity of Service Oriented Architecture methods and capabilities. This pattern will be helpful in the development of net-centric solutions that use cloud computing to support Government agility, lower information technology cost, and overall transformation.

The HCC supports the development of Government cloud computing initiatives with four main architectural tenants:

- ☐ Maintain security and control of personal identifiable information and data on premise (i.e., from cyber-attack, data compromise, and personal identity theft)
- ☐ Obtain agility and cost benefits from the use of public cloud capabilities
- ☐ Develop cloud bursting capability to right-sized private cloud
- ☐ Extend architectural capabilities to mobile and hand-held devices

Participants and Pre-Conditions

Several participants are involved:

- ☐ **Subject Matter Experts** - Bring expertise from field to designing cloud solutions
- ☐ **Standards Developers** -Develop open cloud standards
- ☐ **Cloud Service Providers** - Create and deliver cloud services based on standards

Cloud Service Consumers

Policy Makers - Specify Policies for Governance of Cloud Services

- ☐ Users should agree on the use of appropriate cloud computing delivery and deployment models.
- ☐ There should be an agreed-upon enterprise security, privacy, and confidentiality policy with a complete understanding of data location, access rights, and security methods.
- ☐ There should be means of communication (with sufficient bandwidth, and secured as appropriate) between systems willing to share.
- ☐ Right to share: legal issues and other possible obstacles to information sharing must be resolved to realize the full benefit of the pattern.

- ☐ Will to share: organizational, cultural, and other human issues may create obstacles to information sharing, and should be resolved to realize the full benefit of the pattern
- ☐ Policies and Contracts, including Service Level Agreements (SLAs) MUST be referenced.
- ☐ Each organization/nation can bring customized implementations of the HCC, proposing the cloud(s), cloud services, integration capability, and appropriate cloud standards as part of the implementation.
- ☐ Core services are dependent upon participants/integrators in terms of technology but a Service Oriented Architecture approach is highly recommended.

Structure

Cloud computing is a model for enabling convenient, on-demand network access to a shared pool of configurable computing resources (e.g., networks, servers, storage, applications, and services) that can be rapidly provisioned and released with minimal management effort or service provider interaction.

Three elements critical to the adoption and deployment of cloud computing in government are:

- ☐ Development of robust security, privacy, and confidentiality
- ☐ Acquisition of speed, lower cost, and agility capabilities
- ☐ Extension of enterprise capabilities to end user hand-held devices

The HCC balances the enterprise need for secure, private, and confidential data sharing with the use of public cloud elastic scale and utility-based costing, and extends the enterprise environment to mobile and hand-held devices as the primary cloud computing entry point.

HCC On-Premise Environment

The enterprise on-premise environment is the foundation of the HCC pattern. Integration and utilization of the on-premise environment is critical, as it is generally the source of enterprise secure data. It provides Government agencies a known secure enterprise environment for development of enterprise data and applications, and by using it as part of the cloud computing capability, it reduces the concern for lack of control, security, privacy, and confidentiality of data when developing cloud solutions.

Remember

HCC Public and Private Clouds

Public and private clouds offer speed of acquisition and use, utility-based pricing, and more agility—especially when integrated with the enterprise on-premise environment. Public cloud providers offer infinitely elastic automatic scale and virtual private cloud capabilities that lower cost and seek to manage security risk. Private clouds offer right-sized cloud bursting capabilities for

applications that extend beyond traditional short-lived public cloud needs, or for applications that need to run when bandwidth or survivability issues exist.

Must Do

In order to make use of cloud services part of an agency enterprise solution, interoperability issues, cloud integration, and portability issues need to be resolved. The HCC pattern, with its integration of on-premise enterprise environments with public and private cloud capabilities, balances benefits and risks associated with cloud computing adoption.

HCC Mobile and Hand-Held Devices

This design pattern provides for the extension of enterprise capabilities to multiple mobile and hand-held devices.

HCC Deployment Scenarios

The HCC pattern supports the following notional deployment scenarios:

- ☐ "Cloud bursting" to support cyclic data processing requirements
- ☐ Cloud-based collaboration environments
- ☐ Virtually binding shipboard IT infrastructures to create a battle group Infrastructure as a Service platform
- ☐ Virtually binding land vehicle-based servers and data storage resources to create battlefield data centers
- ☐ Dynamic provisioning of virtual cloud-based servers to support automated exploitation and dissemination of data from unmanned aerial vehicles

Post-conditions

Tips

The HCC pattern offers the ability to provide increased adoption of cloud computing. It provides increased interoperability and portability of cloud-based solutions, and supports nascent cloud computing standards.

- ☐ Implementation Guidance
- ☐ Standards
- ☐ Technology:
  - Service Oriented Architecture technologies
  - Virtualization technologies
  - Cloud computing technologies
  - Enterprise security technologies
  - Mobile hand-held application development technologies

Standards:

Each community of interest can identify a relevant standard. Cloud computing standard, while nascent, will mature over time and guide the pattern implementation. Critical areas for development of cloud standards in support of this pattern include:

- ☐ Federated identity across clouds
- ☐ Metadata and data exchanges among clouds
- ☐ Movement of VMs with context between clouds
- ☐ Portable tools for developing, deploying, and managing cloud applications
- ☐ Standards for describing resource/performance capabilities and requirements
- ☐ Standardized monitoring, auditing, and reporting outputs
- ☐ Common representations (abstract, APIs) for interfacing to cloud resources
- ☐ Cloud-independent representation for describing policies and governance

Expert Advice

Lessons learned

Service Oriented Architecture is the foundational enabler to cloud computing.

Remember

- ☐ Matching of cloud deployment and delivery models drives business/agency successful cloud adoption.
- ☐ Cloud standards are nascent and evolving.
- ☐ Mobile devices are a critical entry point for cloud applications.
- ☐ Identification of appropriate cloud services contributes to a successful business model.
- ☐ Cloud computing is not appropriate for all enterprise application capabilities.
- ☐ Significant attention should be paid to the business and culture in developing the security model.
- ☐ Cloud computing does not obviate the need for standard enterprise architecture, security, and implementation of best practices and methods.

Constraints and Opportunities

- ☐ Real-time access to data from clouds or cloud services
- ☐ Real-time dynamic brokers of clouds or cloud services
- ☐ Certification and accreditation of clouds and cloud services
- ☐ Federated cloud architectures

Legal and Cultural Issues

Support critical missions associated with public outreach, citizen engagement, personnel recruitment/training, and idea generation with greater speed and agility at a lower cost.

The NCOIC HCC is currently under development but is expected to be formally released in late 2011.

STEP TWO. THE NEED

In the past year, cloud computing has morphed from an obscure concept into the driving force behind Federal information technology. Although its merits are still hotly debated, this new approach promises speed, agility, and low cost with greatly improved security, privacy, and confidentiality. When unveiled last year, Federal CIO, Vivek Kundra, highlighted three key elements of this new initiative. The first major element was the Apps.gov site, a clearinghouse for business, social media, and productivity applications, as well as cloud IT services. According to Kundra, the second element of the effort will be budgeting. In FY 2010, cloud computing pilot projects were pushed hard, and followed in 2011 by the issuance of more definitive guidance to agencies throughout government. The final key element will include policy planning and architecture consisting of centralized certifications, target architectures, and formal security, privacy, and procurement processes.

While many still debate the value of cloud computing, deployment of this grand vision will also depend on the realization of cloud interoperability (integration of cloud capabilities), and portability (the ability to move from one cloud service/provider to another). In a world of many and varied cloud computing solutions, government users must be able to freely move from one cloud vendor to another. This is not only an important technical requirement, but is also critical to protecting federal procurement integrity.

Budget Reduction

President Obama FY 2010 Budget

"Of the investments that will involve up-front costs to be recouped in outyear savings, cloud-computing is a prime case in point. The Federal Government will transform its Information Technology Infrastructure by virtualizing data centers, consolidating data centers and operations, and ultimately adopting a cloud-computing business model. Initial pilots conducted in collaboration with Federal agencies will serve as testbeds to demonstrate capabilities, including appropriate security and privacy protection at, or exceeding, current best practices, developing standards, gathering data, and benchmarking costs and performance. The pilots will evolve into migrations of major agency capabilities from agency computing platforms to base agency IT processes and data in the cloud. Expected savings in the outyears, as more agencies reduce their costs of hosting systems in their own data centers, should be many times the original investment in this area," according to the budget description. It will also increase government transparency.

Federal Budget Planning

The President's FY 2011 Budget highlights cloud computing as a major part of the strategy to achieve efficient and effective IT. Federal agencies are to deploy cloud computing solutions to improve the delivery of IT services where the cloud computing solution has demonstrable benefits versus the status quo. OMB, as part of the FY 2011 Budget Process, requested all agencies to evaluate cloud computing alternatives as part of their budget submissions for all major IT investments, where relevant. Specifically:

- ☐ By September 2011 – all newly-planned or performing major IT investments acquisitions must complete an alternatives analysis that includes a cloud computing-based alternative as part of their budget submissions.
- ☐ By September 2012 – all IT investments making enhancements to an existing investment must complete an alternatives analysis that includes a cloud computing-based alternative as part of their budget submissions.
- ☐ By September 2013 – all IT investments in steady-state must complete an alternatives analysis that includes a cloud computing-based alternative as part of their budget submissions.

Increased Efficiency

Presidential Executive Order 13514

i. Federal Leadership in Environmental, Energy, and Economic Performance

ii. **Section 1. Policy.** In order to create a clean energy economy that will increase our Nation's prosperity, promote energy security, protect the interests of taxpayers, and safeguard the health of our environment, the Federal Government must lead by example. It is therefore the policy of the United States that Federal agencies shall increase energy efficiency; measure, report, and reduce their greenhouse gas emissions from direct and indirect activities;

iii. **Sec. 2. Goals for Agencies.** In implementing the policy set forth in section 1 of this order, and preparing and implementing the Strategic Sustainability Performance Plan called for in section 8 of this order, the head of each agency shall:

1. promote electronics stewardship, in particular by:

2. implementing best management practices for energy-efficient management of servers and Federal data centers; and

3. increase IT efficiency

In an important industry contribution, The Open Group has published a white paper on how to build and measure cloud computing return on investment (ROI). Produced by the Cloud Business Artifacts (CBA) project of The Open Group Cloud Computing Work Group, the document:

- ☐ Introduces the main factors affecting ROI from cloud computing, and compares the business development of cloud computing with that of other innovative technologies;
- ☐ Describes the main approaches to building ROI by taking advantage of the benefits that cloud computing provides; and
- ☐ Describes approaches to measuring this ROI absolutely, and in comparison, to traditional approaches to IT, by giving an overview of Cloud Key Performance Indicators (KPIs) and metrics

In presenting their model, business metrics were used to translate indicators of cloud computing capacity-utilization curves into direct and indirect business benefits. The metrics used include:

- ☐ Speed of Cost Reduction;

Figure 2. Speed of Cost Reduction

Leveraging Speed and Cost
TCO
Total Cost of Ownership
Time to Market
Time to Value
Time to Production
TRADITIONAL
Adoption of OPEX based services
Adoption of rapid Dev/Test/Deply Lifecycle
Faster rate of cost reduction
CLOUD
Faster time to cost reduction
Time

☐ Optimizing Ownership Use;

Figure 3. Optimizing Ownership Use

License Costs

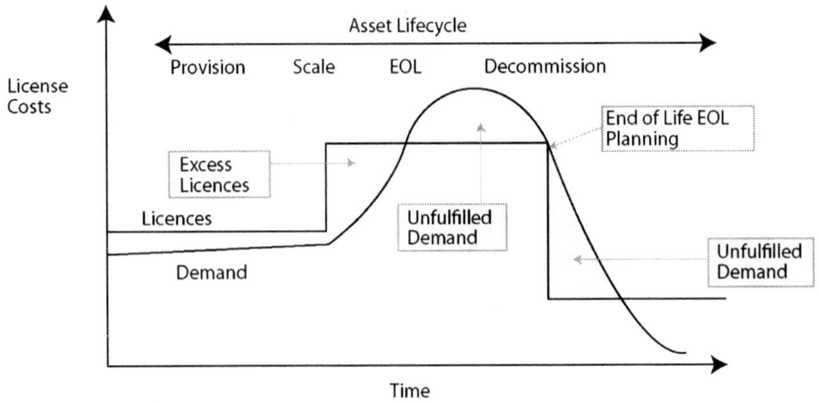

Asset Lifecycle

Provision Scale EOL Decommission

License Costs

End of Life EOL Planning

Excess Licences

Licences

Unfulfilled Demand

Unfulfilled Demand

Demand

Time

☐ Rapid Provisioning;

Figure 4. Rapid Provisioning

Time Compression to Increase and Decrease Deployment

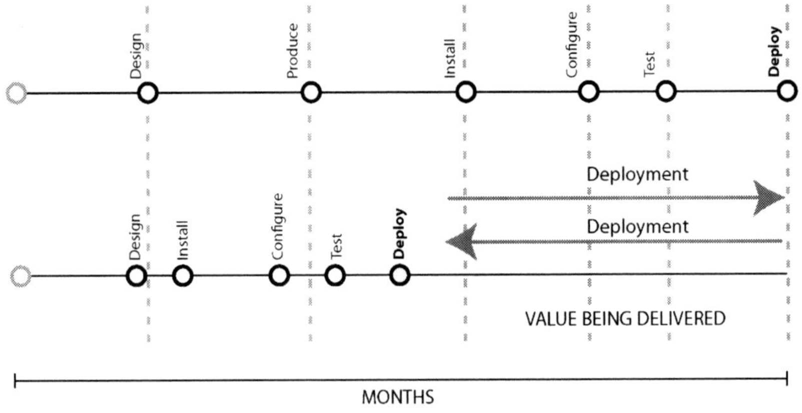

Design Produce Install Configure Test **Deploy**

Deployment

Deployment

Design Install Configure Test **Deploy**

VALUE BEING DELIVERED

MONTHS

☐ Increase Margin;

Figure 5. Increased Margin

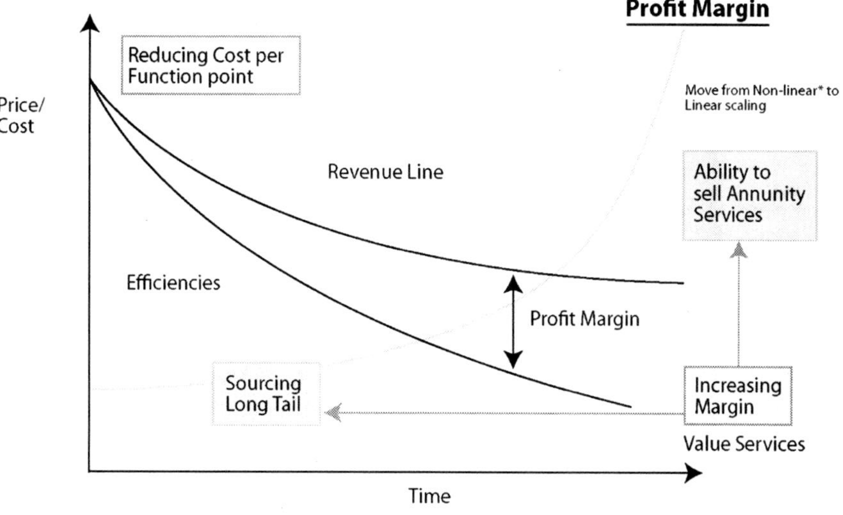

☐ Dynamic Usage

Figure 6. Dynamic Usage

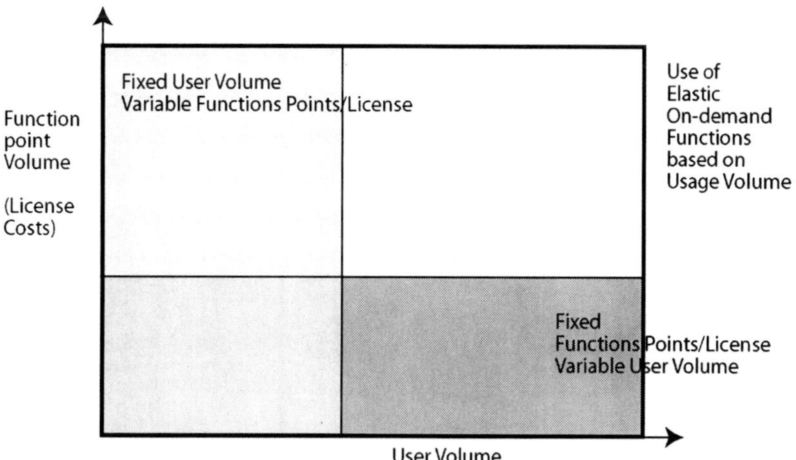

☐ Risk and Compliance Improvement;

Figure 7. Risk and Compliance Improvement

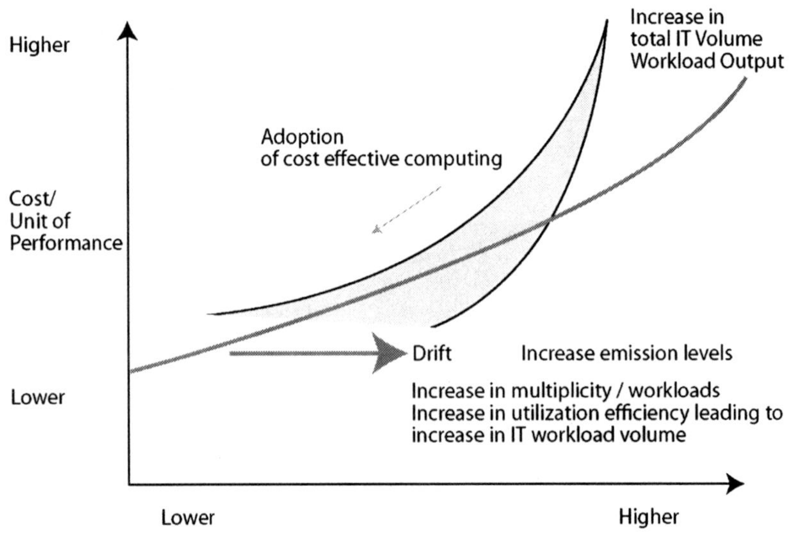

☐ A description of the ROI model used is also provided.

Figure 8. Cloud Computing ROI and KPIs

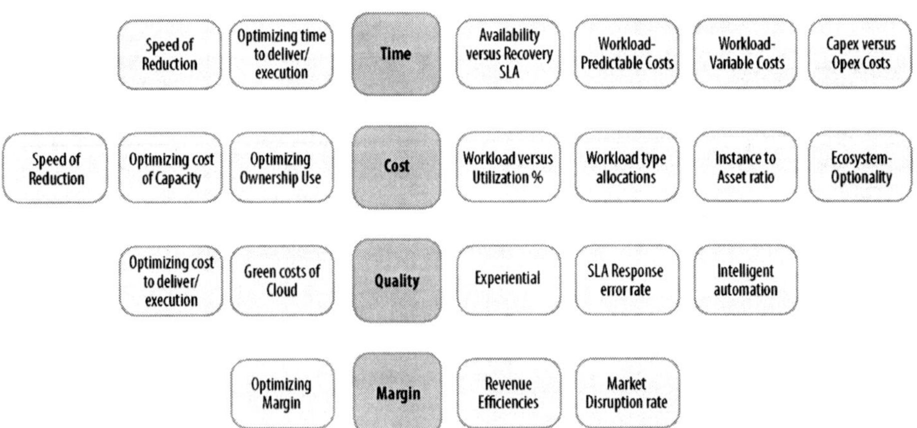

Democratization of Data

President Obama ushered in a new era of government on his first day in office, calling for transparency and open government in his January 21, 2009, memorandum to heads of executive departments and agencies. The memorandum states, "My Administration is committed to creating an unprecedented level of openness in Government. We will work together to ensure the public trust and establish a system of transparency, public participation, and collaboration. Openness will strengthen our democracy and promote efficiency and effectiveness in Government."

On October 19, 2010, GSA announced that federal, state, local, and tribal governments would soon have access to cloud-based Infrastructure as a Service (IaaS) offerings through the government's cloud-based services storefront, Apps.gov. GSA's IaaS contract award allows vendors to provide government entities with cloud storage, virtual machines, and web hosting services to support a continued expansion of governments' IT capabilities into cloud computing environments.

"Offering IaaS on Apps.gov makes sense for the federal government and for the American people. Cloud computing services help to deliver on this Administration's commitment to provide better value for the American taxpayer by making government more efficient," said federal Chief Information Officer, Vivek Kundra. "Cloud solutions not only help to lower the cost of government operations, they also drive innovation across government."

Each year, the government spends tens of billions of dollars on IT products and services, with a heavy focus on maintaining current infrastructure needs and demands. A major element of every federal agency's IT infrastructure includes storage, computing power, and website hosting. New cloud infrastructure offerings can be a way for agencies to realize cost savings, efficiencies, and modernization without having to expend capital resources to expand their existing infrastructure.

On Apps.gov, IaaS offerings will include on-demand self-service that allows government entities to utilize, and discontinue use of, products when, and as, needed. Resource pooling for practically unlimited storage and automatic monitoring of resource utilization are also features. IaaS offerings will also be provided with rapid elasticity for real-time, customizable scaling of service, automatic provisioning of virtual machines, storage, and bandwidth, and visibility into service usage and order management through measured services.

Prior to making IaaS products available through Apps.gov, vendors must complete the Certification and Accreditation (C&A) process at the FISMA Moderate Impact Data security level as administered by GSA. Once granted authority to operate, services will be made available for purchase by government entities through the Apps.gov storefront.

"Through offerings such as IaaS, GSA is providing government entities with easy access to cost-saving, high-value, more efficient technology solutions by doing a major part of the procurement processes upfront," said GSA Associate Administrator of Citizen Services and Innovative Technologies, Dave McClure. "By continuously working with industry, GSA's cloud-based services available through Apps.gov are secure, compliant, and save taxpayer dollars by reducing duplication of security processes across government."

Awarded vendors have assembled skilled teams that will support the development of quality services for government agencies. Awarded vendors and their associated teams include:

- ☐ Apptis Inc. partnered with Amazon Web Services
- ☐ AT&T
- ☐ Autonomic Resources partnered with Carpathia, Enomaly, and Dell
- ☐ CGI Federal Inc.
- ☐ Computer Literacy World partnered with Electrosoft, XCommunications, and Secure Networks
- ☐ NJVC, LLC as a partner to Computer Literacy World
- ☐ Computer Technologies Consultants, Inc., partnered with Softlayer, Inc.
- ☐ Eyak Tech LLC
- ☐ General Dynamics Information Technology partnered with Carpathia
- ☐ Insight Public Sector partnered with Microsoft
- ☐ Savvis Federal Systems
- ☐ Verizon Federal Inc.

Step Three. Setting Goals

What Is the Goal?

Remember

Cloud computing enables convenient, rapid, and on-demand computer network access—most often via the internet--to a shared pool of configurable computing resources (in the form of servers, networks, storage, applications, and services). Quite simply, it is the way computing services are delivered that is revolutionary. Cloud computing allows users to provision computing capabilities rapidly and as needed; that is, to scale out and scale back as required, and to pay only for services used. Users can provision software and infrastructure cloud services on demand with minimal, if any, human intervention. Because cloud computing is based on resource pooling and broad network access, there is a natural economy of scale that can result in lower costs to agencies. In addition, cloud computing offers a varied menu of service models from a private cloud operated solely for one organization, to a public cloud that is available to a large industry group and the general public, and owned by an organization that is selling cloud computing services.

The adoption of safe and secure cloud computing by the Federal government presents an opportunity to close the IT performance gap. Various forms of cloud computing solutions are already being used in the federal government today to save money and improve services.

FCCI Goals

Establish and Manage Governance

Information governance is policy-based management of information designed to lower costs, reduce risk, and ensure compliance with legal, regulatory standards, and/or corporate governance. It governs how information is accessed, secured, and handled throughout the organization, regardless of whether the information resides in paper documents or in encrypted data streams in the cloud.

Information governance policies set the right conditions for people, processes, and technologies to efficiently manage, locate, and deliver information when and where it's needed: What information does my organization have? What information does my organization need to keep? Is information synchronized and consistent? Do the right people have access to the right information at the right time?

The ultimate goal of information governance is to make it faster and easier for organizations to extract actionable insight and value from information in support of their business strategy.

Note: The first Leadership Council for Information Advantage report, *Creating Winning Strategies for Information Advantage*, presented some recommendations for organizations seeking to improve their information governance practices.

Must Do

To maximize the accessibility, utility, and value of corporate information, you need a set of guidelines to ensure information is treated in a consistent way, regardless of whether it resides within traditional enterprise data centers or in clouds. Historically, such guidelines in enterprise environments have been called "information governance."

The basic tenets of enterprise information governance can, and must, be extended to the cloud to help preserve the integrity and usefulness of cloud-based information. If you're just virtualizing your data center to form an internal cloud, the governance changes required will be very minor. For externally hosted private clouds, and public clouds, policies around security, access, compliance, archiving, and information life cycle management do not change: you're still accountable for meeting the same regulatory, privacy, and administrative requirements. However, the controls used to enforce those policies and requirements will change: due diligence in vendor selection, performance and indemnity clauses in SLAs, vendor reports and activity logs, third-party audits, and certifications. These controls may be different from the governance controls organizations had when everything belonged to them and they could physically recover a hard disk if a server went down. But, while these vendor-oriented controls are different, they are not difficult. Managing vendor relationships is a skill every IT department has fine-tuned over the years.

Provide Procurement Leadership

- ☐ Develop acquisition vehicles to ease agency procurement of cloud computing solutions
- ☐ Coordinate across the Federal Acquisition community to ensure adoption and implementation of cloud-related procurement policies and processes
- ☐ Facilitate adoption of the cloud computing storefront

Drive Cloud Technology Innovation

- ☐ Identify common cloud services and foster standards development and security policies
- ☐ Develop architectures that allow agencies to more effectively implement and leverage cloud computing services
- ☐ Establish, manage, and coordinate cloud computing developer communities and application libraries
- ☐ Enable the reuse, modularity and interoperability of cloud computing services

Enable Implementation and Adoption

- ☐ Assist and guide agencies to implement and roll-out cloud solutions (e.g. service provisioning)
- ☐ Facilitate identification of agency partners for pilot activities
- ☐ Develop methodologies to effectively assess and implement services
- ☐ Develop and disseminate cloud services operating and business models

Enable Sustainable and Cost-Effective Computing (Green IT, TCO)

- ☐ Develop and manage business case templates
- ☐ Identify core evaluative criteria
- ☐ Identify cloud computing performance metrics (enterprise and technical)
- ☐ Develop case studies, best practices/lessons learned to specifically demonstrate support of sustainability and cost effective computing

Operate as a Service Provider

- ☐ Identify and offer government-wide services (e-mail, instant messaging, Web 2.0 tools, etc.)
- ☐ Assist agencies in determining their role in developing/hosting services
- ☐ Facilitate common interfaces to integrate existing federal cloud service environments

Conduct Outreach Activities

- ☐ Develop the cloud services communications plan
- ☐ Design and implement a central information portal for federal cloud computing
- ☐ Develop and manage content for communications
- ☐ Manage cloud-related wikis, blogs, portal, and other collaborative media
- ☐ Develop and present training

Additional Goals

- ☐ Reduce the cost of data center hardware, software, and operations
- ☐ Increase the overall IT security posture of the government
- ☐ Shift IT investments to more efficient computing platforms and technologies
- ☐ Promote the use of Green IT by reducing the overall energy and real estate footprint of government data centers

Creating a Cloudsourcing Roadmap for Your Agency

Initial Steps

- ☐ Perform an analysis of what IT services, business applications, or processes you want to deploy in the cloud and identify the optimal cloud service model to support it (e.g., internal cloud, private cloud using external data centers, public cloud).Sequence the services to be deployed over each of the next five years in a roadmap.

Figure 9. Service Model Options and Desired Functions

What functions are you considering putting in a cloud?

	Service Model		
Storage			
Processors			
Application Design			
Application Development			
Application Testing			
Application Deployment			
Application Development			
Collaboration			
Application Security Services			
Application Versioning			
Application Instrumentation			
Web Servers			
Application Servers			
Database Engines			
Application Programming Interfaces (APIs)			
Application Business Process			
Complete Business Application			
Total			
	IaaS	**PaaS**	**SaaS**

Figure 9 can be used as a guide for determining the relative value of each service model option based on the desired cloud-based function.

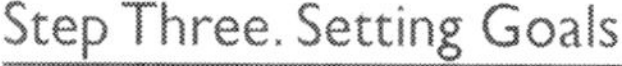

- ☐ Define requirements for each service to be cloud-sourced. Requirements should include service-level metrics such as response time and uptime, as well as security, regulatory, and other information policy restrictions that must be satisfied.
- ☐ Conduct a financial cost/benefits analysis to establish a rationale for why each service should or should not be moved to the cloud. Consumption models should be clear: what is the unit cost component?

#1: Jump into the cloud with a good test case

Organizations have different appetites for risk and different views on what types of applications, services, and information are right to outsource to the cloud. Regardless of their specific concerns and limitations, most organizations should be able to identify a safe way to begin developing their capabilities in managing cloud services. Gaining experience now with cloud services is essential to taking advantage of higher-value future opportunities emerging in the cloud.

Finding a low-risk test case may not be as straightforward as it sounds. Information and IT services that seem easy to deploy in the cloud sometimes are not. Following are four questions to ask when evaluating which IT services lend themselves to "cloud-sourcing," particularly as external private clouds and public clouds.

Can compliance requirements be balanced safely with other priorities? Once you move regulated information to private clouds with external components or to public clouds, the chain of custody for information becomes more complex. Because of this, many IT professionals assume that policy compliance is harder in the cloud—but it is merely a matter of perspective. Even though you have to factor in cloud vendors' logs and attestations in assessing your overall compliance posture (which certainly complicates compliance reporting), you no longer have to bear responsibility for maintaining the IT assets running your services. It is a trade-off.

While failing to meet compliance requirements is never an option, IT leaders need to help their organizations assess new sources of risk, such as chain of custody exposures, introduced by the cloud. Then, once new risk sources are understood, organizations can explore whether compliance obligations may be met in new ways so the organization can reap the benefits of the cloud. Cloud services rarely pose insurmountable challenges for compliance; they simply change the mechanisms for monitoring and enforcement. The real question is: are the perceived benefits of the cloud great enough to justify changes in organizational processes and compliance procedures?

A good illustration of this is e-mail outsourcing. Many organizations today are struggling with whether to move their e-mail to the cloud. Organizations are choosing between hosting their own Exchange servers to meet various e-mail requirements or outsourcing e-mail and learning how to audit e-mail vendors for compliance reporting. For some organizations, the potential cost-savings

and improved service availability of partnering with an outside service provider more than offset the extra steps in compliance assessments.

In addition to chain-of-custody complications, the cloud introduces new compliance challenges in controlling the transmission of regulated information across national boundaries. For instance, banks keeping financial records of Canadian government employees cannot transmit these records or any related correspondence outside of Canada. Many European countries have similar regulations requiring certain information about their citizens be stored only within their national boundaries. Multinational organizations and cloud providers must conform to these requirements by operating their cloud data centers in specific countries. Furthermore, backup and disaster recovery, e-discovery requirements, as well as information accessed by cloud administrators based in foreign jurisdictions must be carefully handled to comply with local regulations. Most cloud providers are familiar with the unique information requirements of the countries where they operate, and they've tailored their cloud services accordingly. For most organizations, addressing the boundary-specific challenges introduced by the cloud is a matter of choosing service providers with experience in regulated jurisdictions, and writing restrictions and penalties for non-conformance into contracts.

Is it an IT function or service your internal organization has mastered? Organizations often are tempted to outsource things they don't fully understand, thinking vendors can provide all the requisite expertise and solutions– like a magic bullet. Looking for outside solutions to poorly defined problems can be especially tempting when time is of overriding importance in getting an IT service up and running. While building from a ready-made cloud platform or outsourcing IT services to cloud vendors can save valuable time, if program requirements aren't clearly understood, costs inevitably balloon and the organization often ends up stuck with an imperfect solution. When assessing which IT assets to move to the cloud, it is best to start off with a function or process your organization already knows inside and out. That way, you recognize the conditions for success, can lay out detailed requirements for what matters, and you know what exactly to look for in prospective vendors.

Can you use a standardized service? Processes or functions that are standardized across companies or business units naturally lend themselves to the cloud. Public cloud services often provide best-in-breed functionality for lower cost. The downside is less customization and control, but that might not be a big deal for many standardized services and functions. Salesforce.com is an outstanding example of this. Its features are well suited to managing customer information and sales processes, whether you're selling paper or professional services. Nevertheless, we're seeing a trend toward configurable public cloud services as cloud providers increasingly offer industry-specific permutations or more granular controls to tailor their offerings to a wider range of customers.

Is the pilot project easily implementable? In scoping projects for public or externally hosted private clouds, if you have to choose between process complexity and technical complexity, choose the latter. In our experience, the technical aspects of cloud adoption—experimenting with APIs, systems integration and service testing—are the easy parts. Business process and service delivery issues tend to be more challenging.

Many IT teams think that just because the application or service they're moving to a cloud isn't "mission critical," the process will be quick and straightforward. This isn't always the case. If your organization hasn't previously worked with a cloud provider and you're setting up a private cloud with custom features, it may take time to delineate what the service provider is responsible for and what your obligations are. Compliance and liability issues can be major sticking points. Defining compliance conditions and establishing liability for intellectual property protection with cloud providers are issues that span well beyond IT. Consequently, they can take a lot of time and coordination to resolve.

Previously, it seemed private cloud service providers offered more expansive liability clauses. Currently, standard liability clauses are more restrained (and arguably realistic), setting limits for financial compensation or stricter conditions on determining fault. IT teams should expect process challenges to occur even in seemingly straightforward projects that aren't handling any regulated or sensitive information.

While it seems process issues should present more of a challenge in public clouds, they often do not. Users typically sign up for services ad hoc, agreeing to strict terms and conditions in which public cloud providers typically promise nothing and assume minimal, if any, liability.

NIST has launched the U.S. Government Cloud Computing Business Use Case Working Group in order to help agencies in the development of cloud compatible use cases. E-mail, geospatial data exchange, and services management are some of the first use cases under development. Appendix 6 provides the NIST Cloud Business Use Case template. Additional information is available online at http://collaborate.nist.gov/twiki-cloud-computing/bin/view/CloudComputing/BusinessUseCasesCall03

Mitre and The OPEN Group have also published guides that can help in this initial analysis. The Cloud Buyers Decision Tree (http://www.opengroup.org/cloud/whitepapers/wp_cloud_dt/execsumm.htm) leads the user through a series of questions based on successful commercial cloud computing implementations. *A Decision Process for Applying Cloud Computing in Federal Environments* by Mitre Corporation (http://www.mitre.org/work/tech_papers/2010/10_1070/) provides a structured engineering process for scoping possible cloud options.

Figure 10. Open Group Cloud Buyer's Decision Tree

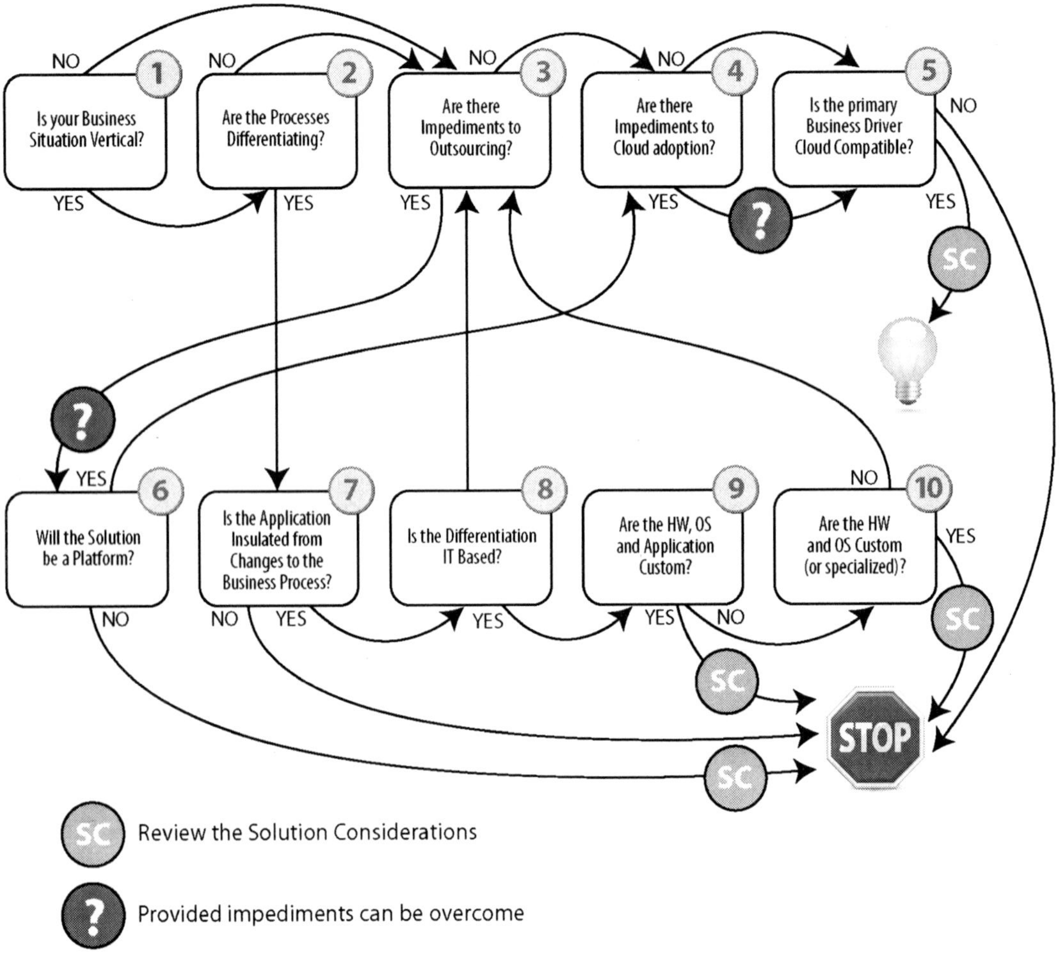

1 Is your Business Situation Vertical?
2 Are the Processes Differentiating?
3 Are there Impediments to Outsourcing?
4 Are there Impediments to Cloud adoption?
5 Is the primary Business Driver Cloud Compatible?
6 Will the Solution be a Platform?
7 Is the Application Insulated from Changes to the Business Process?
8 Is the Differentiation IT Based?
9 Are the HW, OS and Application Custom?
10 Are the HW and OS Custom (or specialized)?

NO YES NO YES NO YES NO YES NO YES YES NO NO YES YES YES NO NO YES

? SC SC SC SC SC STOP

SC Review the Solution Considerations

? Provided impediments can be overcome

Figure 11. Mitre Corporation Cloud Computing Decision Process

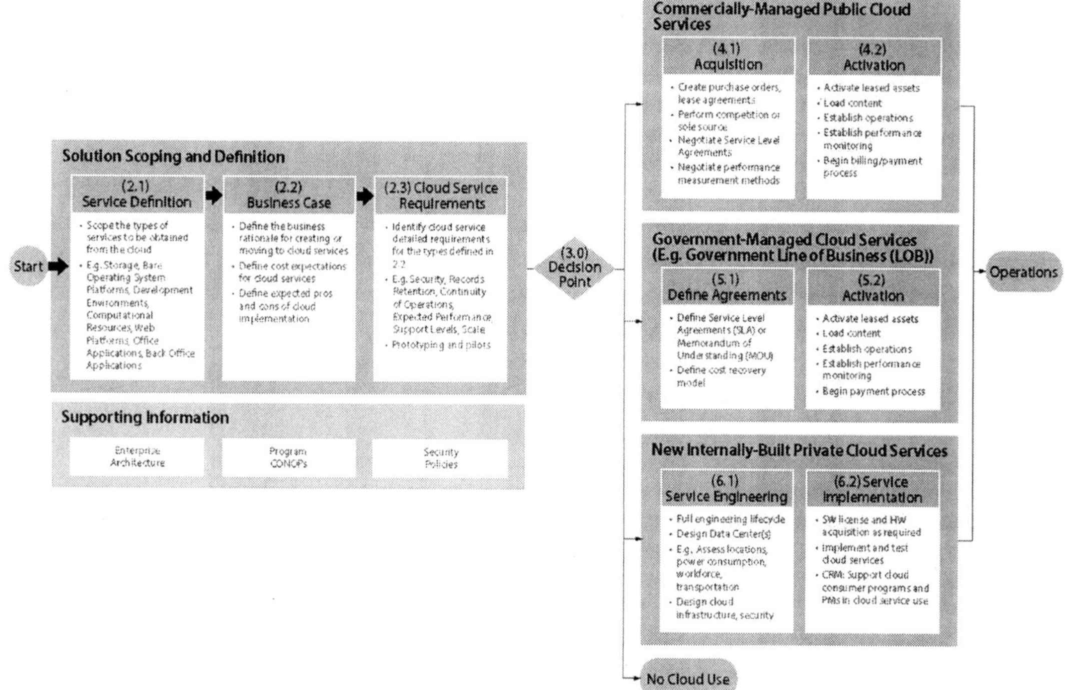

#2: Own the information, even if you own nothing else

Once you know what you're moving to the cloud, assert your organization's right to own the information, even if you don't own the infrastructure, application, or service associated with that information. Externally hosted private cloud and public cloud services may offload the hassles of owning and maintaining expensive IT assets, but they don't offload information liability. If a privacy breach occurs with your consumer information in an external cloud, it's your organization's reputation that suffers.

Because your organization is liable for its information regardless of where it resides, your organization must have the means to manage it appropriately. Ironically, the people who often must be made to understand this are not your cloud vendors—they are your own employees.

It is unrealistic to prevent business units from provisioning their own cloud applications—a niche finance application here, a public cloud CRM service there—but the organization must institute policies and procedures to ensure the organization has the ability to monitor how cloud-based information is managed and, if needed, bring in that information for use within the enterprise.

Remember

Any new public cloud service that employees activate will inevitably involve the IT department at some point. When that happens, the organization's information governance policies should kick in. If the new service handles sensitive or regulated information is it clear that the organization has done the due diligence and put the contractual terms in place to safeguard that information? Has the organization trained employees on how to handle such information in cloud environments?

How is backup handled by the service provider, and is the information important enough to be duplicated within the enterprise? Because it is often hard to forecast future uses for information, err on the side of caution by making sure cloud-based data and content can be brought back into the enterprise, even if you ultimately choose not to keep it.

#3: Don't take terminology for granted

Tips

In working with cloud service providers, it is often easy to overlook basic things like whether you define important terminology in the same way. Even though policies and regulations often detail how information should be handled, there's always room for interpretation as to how to comply. For instance, Rule 17a-4 of the Securities Exchange Act requires instant messages and e-mails to be kept "in an easily accessible place" for two years. Your e-mail cloud provider might have a very different interpretation of "easily accessible" than you do.

Here is a simple hypothetical situation, if a financial institution is asked to disclose all its dealings with a particular client within the last four years, will the search interfaces and message retrieval functions from the firm's e-mail provider be sufficient to support e-discovery and related report generation? Or will the cloud provider dump four years of raw data into that company's lap and let them figure out how to sort and correlate the information?

Organizations should review their information governance policies to identify the elements of highest risk and establish authoritative definitions for key vocabulary in those areas. This should be done not only to ensure vendors' expectations are aligned with your own, but also to help your internal organization focus on what is absolutely essential to control. Once the definitions and key control points are identified, it makes it easier to figure out what IT assets need to be allocated or reconfigured to ensure everyone and everything performs to specifications, outsourced or not.

#4: Hope for standards, but prepare to integrate

We would all like assurance that cloud investments made today will interoperate with other clouds, now and in the future. But in reality, the cloud simply isn't mature enough to have developed standard specifications for platform interoperability and data exchange. XML as a data format is a useful carry-over from the web, but it is not compatible with legacy data formats used

in many enterprise back-end systems. Various proposals for interoperability standards have been put forth for cloud components such as virtual machine metadata, service provider APIs, and identity management, but these standards are still in their early phases.

In the meantime, organizations entering the cloud do so with some risk. If they invest in porting a lot of their information and applications to a particular cloud vendor's API, they risk vendor lock-in. Staying out of the cloud probably isn't a realistic option either, since it's so easy for users with a credit card and a perceived business need to provision their own cloud services. Just as the web compelled us to integrate e-commerce systems to legacy ERP, and open source spawned a mass porting of code from proprietary UNIX, IT teams must proactively anticipate and plan for the next stage of information integration in the cloud.

The trick is to lay the strategic groundwork early for data integration down the road. Organizations should insist their cloud vendors provide clear documentation on the data formats and schemas used to store information in their systems. Preferably, organizations should keep information architecture at the top of their minds as they select cloud service providers.

#5: Control cloud platform proliferation

In the absence of cloud interoperability standards, the best way to limit information fragmentation—and head off a huge integration effort years down the road—is to try to minimize the number of different cloud platforms that require support. Admittedly, this is much easier said than done. Business groups need the flexibility to choose IT tools that support their work, and IT needs to support the business's ability to differentiate and gain a competitive edge. Rarely does that involve standard tools.

But to the extent possible, IT teams should help business units look for shared requirements in standardized business functions such as finance, HR, and CRM. Then, teams can identify cloud platforms that meet these various needs and consolidate the organization's services on them wherever possible. This should not only improve your organization's ability to share information across business units, it should also result in greater negotiating leverage for favorable contract terms and pricing, as you'll be offering up a bigger piece of business.

Standardizing information platforms is very hard to do without buy-in at the top levels of the organization. It is up to the CIO and other business leaders to prioritize information architecture in the selection of new IT systems and services—including cloud platforms—and to support measures to maximize the utility of information within their organizations. With these goals in mind, it is crucial for IT leaders and business executives to collaboratively develop a strategic plan for how the organization will make use of the cloud.

#6: Make your information "cloud ready"

Remember

The process of information integration is often much harder than the technology of data integration. It can be very difficult to muster the organizational will to tackle information integration projects. In some cases, it's hard to prove potential ROI. In other cases, data integration projects are hard to justify in the face of competing priorities.

However, organizations that have organized their data sets well enough to use them across multiple platforms will be best positioned to take full advantage of cloud services. They will be able to migrate enterprise information to cloud services more easily. Organizations without well-integrated information will likely find themselves confined to infrastructure-level cloud services such as computing and storage. In the cloud, the companies that have already figured out how to use their information for business advantage will quickly widen the gap between themselves and the companies still struggling to make sense of their data and content.

This is why it is essential for organizations to redouble their efforts to integrate their information and make it "cloud ready." This means getting into the habit of encrypting data, a practice that will become especially important as more of your corporate information travels through externally operated resources in the cloud. It also means redoubling efforts to tag fixed data and consolidate storage repositories. Transformation technologies such as ETL tools (extract, transform, load) can simplify the conversion of data from one format to another. The goal should be to convert information into one common format—probably XML, the gold-standard for the cloud. Information preserved as XML becomes more portable and broadly searchable, qualities that remain intact as you migrate services to the cloud, even if XML schemas differ between cloud solutions.

Furthermore, organizations should consider decommissioning outdated or underutilized legacy applications. Research from various IT research firms estimate that 10-20 percent of overall IT budgets support underutilized or aging applications. That is a lot of money spent on shelfware. Decommissioning such applications not only creates opportunities to move data to XML, which can be more broadly used, it also can save organizations millions of dollars in system consolidation. Lastly, companies should insist that their technology partners demonstrate a deep understanding of these growing requirements and are equally passionate about making their own products and services cloud ready.

#7: Master solution integration

As fewer IT assets reside in internal data centers, technology professionals need to shift their focus from owning and operating IT systems to becoming master information service integrators. Along with linking legacy databases to SaaS, IT teams will also need to connect their various private and public clouds, creating a seamless service environment that works like a single cloud custom-made for the enterprise. This transcends solutions integration. It requires greater synchronization between IT departments and their vendors. It also requires IT managers to act less like a part of the service delivery chain and more like an orchestrator of the entire service experience.

In a cloud computing-oriented world, service integration will become one of the corporate IT teams' most important roles and responsibilities. By focusing on activities that add value to the business, as opposed to simply managing an operational infrastructure, IT can truly become a partner to business.

Requirements:

- ☐ Manage cloud computing executive steering committee and cloud computing agency advisory council
- ☐ Establish and manage communities of practice and working groups
- ☐ Coordinate policy and strategy development and participation in other governance bodies related to cloud computing activities (e.g. security, records management, e-discovery)
- ☐ Establish a Federal Certification and Accreditation process

Examples

U.S. Army

The Department of the Army Experience Center in Philadelphia is piloting the use of a customer relationship management (CRM) tool. The Center is a recruiting center that reaches out to young people who are interested in joining the armed forces. The Center wants to move to real-time recruiting, and to use tools and techniques that are familiar and appeal to its young demographic. They are using a CRM provided by SalesForce.com to track recruits as they work with the Center. Since the tool integrates directly with e-mail, Twitter, and Facebook, recruiters can maintain connections with potential candidates directly after they leave the Center. The Army estimated that to implement a traditional CRM would have cost $500,000. The cloud-based solution has been implemented at the cost of $54,000.

Department of Energy

The Department of Energy is evaluating the cost and efficiencies resulting from leveraging a cloud computing solution across the enterprise to support business and scientific services. The Lawrence Berkeley Lab has deployed over 5,000 mailboxes on Google Federal Premiere Apps and they are now evaluating the use of Amazon Elastic Compute Cloud (EC2) to handle excess

capacity for computers during peak demand. The lab estimates that they will save $1.5 million over the next five years in hardware, software, and labor costs from the deployments they have made.

General Services Administration

Remember

GSA has moved the primary information portal, USA.gov, to a cloud-based host. This enabled the site to deliver a consistent level of access to information as new databases are added, peak usage periods are encountered, and as the site evolves to encompass more services. By moving to a cloud, GSA was able to reduce site upgrade time from nine months to one day, monthly downtime improved from two hours to 99.9% availability, and the GSA realized savings of $1.7M in hosting services. In addition to improved services, cloud computing will be a major factor in reducing the environmental impact of technology and will help achieve important sustainability goals. Effective use of cloud computing can be part of an overall strategy to reduce the need for multiple data centers and the energy they consume. Currently, GSA is supporting OMB in working with agencies to develop plans to consolidate their data centers. Using the right deployment model – private cloud, community cloud, public cloud, or a hybrid model – can help agencies buy improved services at a lower cost within acceptable risk levels, without having to maintain expensive, separate, independent, and often needlessly redundant brick and mortar data centers.

Step Four. Implementation

What Are the Options?

Status Quo

This is not a viable option because budget pressures continue to drive the need to reduce IT expenditures.

Reduce Reliance on IT

Reducing reliance on IT is not a viable option because of the growth of information and the digital universe.

Between now and 2020, the amount of digital information created and replicated in the world will grow to an almost inconceivable 35 trillion gigabytes as all major forms of media – voice, TV, radio, and print – complete the journey from analog to digital.

At the same time, the influx of consumer technologies into the workplace will create stresses and strains on the organizations that must manage, store, protect, and dispose of all this electronic content.

- ☐ In 2009, despite the global recession, the Digital Universe—everything that is sent and saved digitally—set a record. It grew by 62% to nearly 800,000 petabytes. A petabyte is a million gigabytes. Picture a stack of DVDs reaching from the earth to the moon and back.
- ☐ Last year, the Digital Universe grew almost as fast to 1.2 million petabytes, or 1.2 zettabytes.
- ☐ This explosive growth means that by 2020, our Digital Universe will be 44 times as large as it was in 2009. Our stack of DVDs would now reach halfway to Mars.

Source. 2010 Digital Universe Study.

Key Operational Questions

How will we find the information we need when we need it?

We will need new search and discovery tools. Most of the Digital Universe is unstructured data (for example, images and voice packets). We will need new ways to add structure to unstructured data, to look INSIDE the information containers and recognize content such as a face in a security video. In fact, the fastest-growing category in the Digital Universe is metadata, or data about data.

How will we know what information we need to keep, and how will we keep it?

Because there are new technologies tied to storage, we will also need new ways to manage our information. We will need to classify it by importance, know when to delete it, and predict which information we will need in a hurry.

How will we follow the growing number of government and industry rules about retaining records, tracking transactions, and ensuring information privacy?

Compliance with regulations has become an entire industry – a $46 billion industry last year. Many observers, however, feel that still more needs to be done.

How will we protect the information we need to protect?

If the amount of information in the Digital Universe is growing at approximately 50% a year, the subset of information that needs to be secured is growing almost twice as fast. The amount of unprotected, yet sensitive, data is growing even faster.

How Will I Use Cloud Computing?

End User to Cloud

In this scenario, an end user is accessing data or applications in the cloud. Common applications of this type include e-mail hosting and social networking sites. A user of Gmail, Facebook, or LinkedIn accesses the application, and their data, through any browser on any device. Most users do not want to keep up with anything more than a password. Their data is stored and managed in the cloud. Most importantly, the user has no idea how the underlying architecture works. If they can get to the internet, they can get to their data.

Requirements

- ☐ **Identity:** The cloud service must authenticate the end user.
- ☐ **An open client:** Access to the cloud service should not require a particular platform or technology.
- ☐ **Security:** Security (including privacy) is a common requirement to all use cases, although the details of those requirements will vary widely from one use case to the next.
- ☐ **SLAs:** Although service level agreements for end users will usually be much simpler than those for enterprises, cloud vendors must be clear about what guarantees of service they provide.

Enterprise to Cloud to End User

In this scenario, an enterprise is using the cloud to deliver data and services to the end user. When the end user interacts with the enterprise, the enterprise accesses the cloud to retrieve data

and/or manipulate it, sending the results to the end user. The end user can be someone within the enterprise or an external customer.

Requirements

- ☐ **Identity:** The cloud service must authenticate the end user.
- ☐ **An open client:** Access to the cloud service should not require a particular platform or technology.
- ☐ **Federated identity**: In addition to basic the identity needed by an end user, an enterprise user is likely to have an identity with the enterprise. The ideal is that the enterprise user manages a single ID, with an infrastructure federating other identities that might be required by cloud services.
- ☐ **Location awareness:** Depending on the kind of data the enterprise is managing on the user's behalf, there might be legal restrictions on the location of the physical server where the data is stored. Although this violates the cloud computing ideal—that the user should not have to know details of the physical infrastructure—this requirement is essential. Many applications cannot be moved to the cloud until cloud vendors provide an API for determining the location of the physical hardware that delivers the cloud service.
- ☐ **Metering and monitoring:** All cloud services must be metered and monitored for cost control, chargebacks, and provisioning.
- ☐ **Management and Governance:** Public cloud providers make it very easy to open an account and begin using cloud services. This ease-of-use creates the risk that individuals in an enterprise will use cloud services on their own initiative. Management of VMs and of cloud services such as storage, databases, and message queues is needed to track what services are used.
- ☐ **Governance:** This is crucial to ensure that policies and government regulations are followed wherever cloud computing is used. Other governance requirements will be industry- and geography-specific.
- ☐ **Security:** Any use case involving an enterprise will have more sophisticated security requirements than one involving a single end user. Similarly, the more advanced enterprise use cases to follow will have equally more advanced security requirements.
- ☐ **A Common File Format for VMs**: A VM created for one cloud vendor's platform should be portable to another vendor's platform.
- ☐ **Common APIs for Cloud Storage and Middleware:** The enterprise use cases require common APIs for access to cloud storage services, cloud databases, and other cloud middleware services such as message queues. Writing custom code that works only for

a particular vendor's cloud service locks the enterprise into that vendor's system and eliminates some of the financial benefits and flexibility that cloud computing provides.

- ☐ **Data and Application Federation:** Enterprise applications need to combine data from multiple cloud-based sources, and they need to coordinate the activities of applications running in different clouds.
- ☐ **SLAs and Benchmarks:** In addition to the basic SLAs required by end users, enterprises who sign contracts based on SLAs will need a standard way of benchmarking performance. There must be an unambiguous way of defining what a cloud provider will deliver, and there must be an unambiguous way of measuring what was actually delivered.
- ☐ **Lifecycle Management:** Enterprises must be able to manage the lifecycle of applications and documents. This requirement includes versioning of applications and the retention and destruction of data. Discovery is a major issue for many organizations. There are substantial legal liabilities if certain data is no longer available. In addition to data retention, in some cases an enterprise will want to make sure data is destroyed at some point.

Enterprise to Cloud

This use case involves an enterprise using cloud services for its internal processes. This might be the most common use case in the early stages of cloud computing because it gives the enterprise the most control.

In this scenario, the enterprise uses cloud services to supplement the resources it needs:

- ☐ Using cloud storage for backups or storage of seldom-used data
- ☐ Using virtual machines in the cloud to bring additional processors online to handle peak loads (and, of course, shutting down those VMs when they are no longer needed)
- ☐ Using applications in the cloud (SaaS) for certain enterprise functions (e-mail, calendaring, CRM, etc.)
- ☐ Using cloud databases as part of an application's processing. This could be extremely useful for sharing that database with partners, government agencies, etc.

Requirements

The basic requirements of the Enterprise to Cloud use case are much the same as those for the Enterprise to Cloud to End User use case. An open client, federated identity, location awareness, metering and monitoring, management and governance, security, a common file format for VMs, common APIs for cloud storage and middleware, data and application federation, SLAs, and lifecycle management all apply.

Other requirements for this use case are:

- ☐ **Deployment:** It should be simple to build a VM image and deploy it to the cloud as necessary. When that VM image is built, it should be possible to move that image from one cloud provider to another. Deployment of applications to the cloud should be straightforward, as well.
- ☐ **Industry-specific standards and protocols:** Many cloud computing solutions between enterprises will use existing standards such as RosettaNet or OASGIS. The applicable standards will vary from one application to the next, and from one industry to the next.

Enterprise to Cloud to Enterprise

This use case involves two enterprises using the same cloud. The focus here is hosting resources in the cloud so that applications from the enterprises can interoperate. A supply chain is the most obvious example for this use case.

Requirements

The basic requirements of the Enterprise to Cloud to Enterprise use case are much the same as those for the Enterprise to Cloud use case. Identity, an open client, federated identity, location awareness, metering and monitoring, management and governance, security, industry-specific standards, common APIs for storage and middleware, data and application federation, SLAs and lifecycle management all apply.

Other requirements for this use case are:

- ☐ **Transactions and concurrency:** For applications and data shared by different enterprises, transactions and concurrency are vital. If two enterprises are using the same cloud-hosted application, VM, middleware or storage, it is important that any changes made by either enterprise are done reliably.
- ☐ **Interoperability:** Because more than one enterprise is involved, interoperability between the enterprises is essential.

Private Cloud

The Private Cloud use case is different from the others in that the cloud is contained within the enterprise. This is useful for larger enterprises. For example, if the payroll department has a surge in workload on the 15th and 30th of each month, they need enough computing power to handle the maximum workload, even though their everyday workload for the rest of the month is much lower. With a private cloud, computing power is spread across the enterprise. The payroll department gets extra cycles when they need it and other departments get extra cycles when they need it. This can deliver significant savings across the enterprise.

Remember

Requirements

The basic requirements of the Private Cloud use case are an open client, metering and monitoring, management and governance, security, deployment, interoperability, a common VM format, and SLAs.

Note that a private cloud does not require identity, federated identity, location awareness, transactions, industry standards, common APIs for cloud middleware and lifecycle management. In many cases, consumers have to use a private cloud so that location awareness will no longer be an issue. Keeping the cloud inside the enterprise removes many of the requirements for identity management, standards, and common APIs.

Changing Cloud Vendors

This use case involves working with a different cloud vendor, either adding an additional vendor or replacing an existing one. It applies to all of the other use cases discussed in this handbook. Being able to work with other vendors without major changes is one of the main benefits of openness and standardization.

There are four different scenarios here, each of which has slightly different requirements. In general, changing cloud vendors requires an open client, location awareness, security, SLAs, a common file format for VMs and common APIs for cloud storage and middleware.

Scenario 1: Changing SaaS vendors

In this scenario, a cloud customer changes SaaS vendors. Both SaaS vendors provide the same application (CRM, accounting, word processing, etc.). Documents and data created with one vendor's software should be importable by the second vendor's software. In some cases, the customer might need to use the two vendors interchangeably.

Requirements

- ☐ **Industry-specific standards:** Moving documents and data from one vendor's application to another requires both applications to support common formats. The formats involved will depend on the type of application.

In some cases, standard APIs for different application types will also be required.

It is important to note that there is nothing cloud-specific to these requirements. The standards for moving a document from Zoho to Google Docs, are the same standards for moving a document from Microsoft Office to OpenOffice.

Scenario 2: Changing middleware vendors

In this scenario, a cloud customer changes cloud middleware vendors. Existing data, queries, message queues, and applications must be exportable from one vendor and importable by the other.

Requirements

- ☐ **Industry-specific standards:** Moving documents and data from one vendor's middleware to another requires both applications to support common formats. The formats involved will depend on the type of application.
- ☐ **Common APIs for Cloud Middleware:** This includes all of the operations supported by today's cloud services, including cloud databases, cloud message queues, and other middleware. APIs for connecting to, creating, and dropping databases and tables.

Cloud database vendors have enforced certain restrictions to make their products more scalable and to limit the possibility of queries against large data sets taking significant resources to process. For example, some cloud databases don't allow joins across tables, and some don't support a true database schema. Those restrictions are a major challenge to moving between cloud database vendors, especially for applications built on a true relational model.

Other middleware services such as message queues are more similar, so finding common ground among them should be simpler.

Scenario 3: Changing cloud storage vendors

In this scenario, a cloud customer changes cloud storage vendors.

Requirements

- ☐ **A Common API for Cloud Storage:** Code that reads or writes data in one cloud storage system should work with a different system with as few changes as possible, and those changes should be confined to configuration code. In a JDBC application, as an example, the format of the URL and the driver name are different for different database vendors, but the code to interact with the database is identical.

Scenario 4: Changing VM hosts

In this scenario, a cloud customer wants to take virtual machines built on one cloud vendor's system and run it on another cloud vendor's system.

Requirements.

- ☐ **A common format for virtual machines:** The VM format should work with any operating system.

The assumption here is that the virtual machines themselves are running an operating system such as Windows or Linux. This means that the user of the virtual machine has chosen a platform prior to building a VM for the cloud, so there are no cloud-specific requirements for the software running inside the VM.

Hybrid Cloud

This use case involves multiple clouds working together, including both public and private clouds. A hybrid cloud can be delivered by a federated cloud provider that combines its own resources with those of other providers. A broker can also deliver a hybrid cloud—the difference is that a broker does not have any cloud resources of its own. The provider of the hybrid cloud must manage cloud resources based on the consumer's terms.

It is important to note that, to the consumer of a hybrid cloud, this use case is no different from the End User to Cloud use case discussed earlier. The user has no knowledge of what the hybrid cloud provider actually does.

Requirements

- ☐ All of the requirements of the previous use cases (except Transactions and concurrency) apply here, particularly Security, Data and Application Federation and Interoperability.
- ☐ **SLAs:** A machine readable, standard format for expressing an SLA. This allows the hybrid cloud provider to select resources according to the consumer's terms without human intervention.

The requirements for a community cloud are a subset of the requirements for the hybrid cloud. A community cloud has an infrastructure shared among enterprises with a common purpose.

Government Cloud Governance

Cloud computing systems are hosted on large, multi-tenant infrastructures. This shared infrastructure provides the same boundaries and security protocols for each customer. In such an environment, completing the security assessment and authorization process separately by each customer is redundant. Instead, a government-wide risk and authorization program would enable providers, and the program office, to complete the security assessment and authorization process one time and share the results with customer agencies.

Additionally, the Federal Information Security Management Act (FISMA) and NIST special publications provide Federal Agencies with guidance and framework needed to securely use cloud systems. However, interpretation and application of FISMA requirements and NIST Standards vary greatly from agency to agency. Not only do agencies have varying numbers of security requirements at or above the NIST baseline, many times additional requirements from multiple

agencies are not compatible on the same system. A government-wide risk and authorization program for cloud computing would allow agencies to completely leverage the work of an already completed authorization or only require an agency to complete delta requirements (i.e. unique requirements for that individual agency).

Finally, security authorizations have become increasingly time-consuming and costly both for the Federal Government and private industry. A government-wide risk and authorization program will promote faster and cost effective acquisitions of cloud computing systems by using an 'approve once and use often' approach to leveraging security authorizations. Additionally, such a program will promote the Administration's goal of openness and transparency in government. All of the security requirements, processes, and templates will have to be made publicly available for consumption not only by Federal agencies, but by private vendors as well. This will allow Federal agencies to leverage this work at their agency, and allow private industry to finally have the full picture of what a security authorization will entail prior to being in a contractual relationship with an agency.

Federal Risk and Authorization Management Program (FedRAMP)

Figure 12. FedRAMP Governance Model

Federal CIO Council
Responsible for setting priorities, Providing strategic guidance, and ensuring that program objectives are clearly communicated to Federal Agencies.

Federal CIO
Provides overall direction and program oversight. Responsible for program performance and accountability.

ISIMC
Responsible for socializing and reviewing FedRAMP documents, vetting cloud computing best practices, lessons learned, emerging concepts, etc.

FedRAMP
JAB | JAB TRs | Operations
Responsible for developing and maintaining FedRAMP security requirements, reviewing assessments, authorizing cloud computing solutions.

The Federal Risk and Authorization Management Program (FedRAMP) has been established to provide a standard approach to Assessing and Authorizing (A&A) cloud computing services and products. FedRAMP allows joint authorizations and continuous security monitoring services for Government and commercial cloud computing systems intended for multi-agency use. Joint authorization of cloud providers results in a common security risk model that can be leveraged across the Federal Government. The use of this common security risk model provides a consistent baseline for cloud-based technologies. This common baseline ensures that the benefits of cloud-

based technologies are effectively integrated across the various cloud computing solutions currently proposed within the government. The risk model will also enable the government to 'approve once and use often' by ensuring multiple agencies gain the benefit and insight of the FedRAMP's Authorization and access to service providers' authorization packages.

The objective of FedRAMP is threefold:

- ☐ Ensure that information systems/services used government-wide have adequate information security;
- ☐ Eliminate duplication of effort and reduce risk management costs; and
- ☐ Enable rapid and cost-effective procurement of information systems/services for Federal agencies.

By providing unified government-wide risk management for enterprise-level IT systems, FedRAMP will enable agencies to either use or leverage authorizations with:

- ☐ An inter-agency vetted approach;
- ☐ Consistent application of Federal security requirements;
- ☐ Consolidated risk management; and
- ☐ Increased effectiveness and management cost savings.

Remember

The ability to embrace cloud computing capabilities for federal departments and agencies brings advantages and opportunities for increased efficiencies, cost savings, and green computing technologies. However, cloud computing also brings new risks and challenges in securely using cloud computing capabilities as good stewards of government data. In order to address these concerns, the U.S. Chief Information Officer (U.S. CIO) requested the Federal CIO Council launch a government-wide risk and authorization management program. The government-wide Federal Risk and Authorization Management Program (FedRAMP) aims to provide joint security assessment, authorizations, and continuous monitoring of cloud computing services for all Federal Agencies to leverage.

Cloud computing is not a single capability, but a collection of essential characteristics that are manifested through various types of technology deployment and service models. A wide range of technologies fall under the title of "cloud computing," and the complexity of their various implementations may result in confusion among program managers. The guidelines represent a subset of the National Institute of Standards and Technology (NIST) definition of cloud computing, with three service models; Software as a Service, Platform as a Service, and Infrastructure as a Service (SaaS, PaaS, and IaaS).

The decision to embrace cloud computing technology is a risk-based decision, not a technology-based decision. As such, this decision, from a risk management perspective, requires inputs from all stakeholders, including the CIO, CISO, Office of General Counsel (OGC), privacy official, and the program owner. Once the business decision has been made to move towards a

cloud computing environment, agencies must then determine the appropriate manner for their security assessments and authorizations.

In addition, FedRAMP will work in collaboration with the CIO Council and Information Security and Identity Management Committee (ISIMC) to constantly refine and keep documentation up-to-date with cloud computing security best practices. Separate from FedRAMP, ISIMC has developed guidance for agency use on the secure use of cloud computing in *Federal Security Guidelines for Cloud Computing.*

Transparent Path For Secure Adoption Of Cloud Computing

The security guidance and FedRAMP assessment and authorization process aims to develop robust cloud security governance for the Federal Government. It represents collaboration amongst security experts and representatives throughout government including all of the CIO Council Agencies. By following these requirements and processes, Federal agencies will be able to take advantage of cloud-based solutions to provide more efficient and secure IT solutions when delivering products and services to its customers.

Continuous Monitoring

A critical aspect of managing risk to information from the operation and use of information systems involves the continuous monitoring of the security controls employed within, or inherited by, the system. Conducting a thorough point-in-time assessment of the deployed security controls is a necessary, but not sufficient, condition to demonstrate security due diligence. An effective organizational information security program also includes a rigorous continuous monitoring program integrated into the System Development Life Cycle (SDLC). The objective of the continuous monitoring program is to determine if the set of deployed security controls continue to be effective over time in light of the inevitable changes that occur. Continuous monitoring is a proven technique to address the security impacts on an information system resulting from changes to the hardware, software, firmware, or operational environment. A well-designed and well-managed continuous monitoring program can effectively transform an otherwise static security control assessment and risk determination process into a dynamic process that provides essential, near real-time security status-related information to organizational officials in order to take appropriate risk mitigation actions and make cost-effective, risk-based decisions regarding the operation of the information system. Continuous monitoring programs provide organizations with an effective mechanism to update Security Plans, Security Assessment Reports, and Plans of Action and Milestones (POA&Ms)

An effective continuous monitoring program includes:

- ☐ Configuration management and control processes for information systems;
- ☐ Security impact analyses on proposed or actual changes to information systems and environments of operation;
- ☐ Assessment of selected security controls (including system-specific, hybrid, and common controls) based on the defined continuous monitoring strategy;

☐ Security status reporting to appropriate officials; and

☐ Active involvement by authorizing officials in the ongoing management of information system-related security risks.

Must Do

A service provider is required to develop a strategy and implement a program for the continuous monitoring of security control effectiveness, including the potential need to change or supplement the control set, taking into account any proposed/actual changes to the information system or its environment of operation. Continuous monitoring is integrated into the organization's system development life cycle processes. Robust continuous monitoring requires the active involvement of information system owners and common control providers, chief information officers, senior security officers, and authorizing officials. Continuous monitoring allows an organization to: (i) track the security state of an information system on a continuous basis; and (ii) maintain the security authorization for the system over time in highly dynamic environments of operation with changing threats, vulnerabilities, technologies, and missions/business processes. Continuous monitoring of security controls using automated support tools facilitates near real-time risk management and represents a significant change in the way security authorization activities have been employed in the past. Near real-time risk management of information systems can be accomplished by employing automated support tools to execute various steps in the Risk Management Framework including authorization-related activities. In addition to vulnerability scanning tools, system and network monitoring tools, and other automated support tools that can help to determine the security state of an information system, organizations can employ automated security management and reporting tools to update key documents in the authorization package including the security plan, security assessment report, and plan of action and milestones. The documents in the authorization package are considered "living documents" and updated accordingly based on actual events that may affect the security state of the information system.

Continuous Monitoring Requirements

FedRAMP is designed to facilitate a more streamlined approach and methodology to continuous monitoring. Accordingly, service providers must demonstrate their ability to perform routine tasks on a specifically-defined, scheduled basis to monitor the cyber security posture of the defined IT security boundary. While FedRAMP will not prescribe specific toolsets to perform these functions, FedRAMP does prescribe their minimum capabilities. Furthermore, FedRAMP will prescribe specific reporting criteria that service providers can utilize to maximize their FISMA reporting responsibilities while minimizing the resource strain that is often experienced.

Reporting and Continuous Monitoring

Remember

Maintenance of the security Authority To Operate (ATO) will be through continuous monitoring of security controls of the service providers system and its environment of operation to determine if the security controls in the information system continue to be effective over time in light of changes that occur in the system and environment. Through continuous monitoring, security controls, and supporting deliverables, are updated and submitted to FedRAMP. The submitted deliverables provide a current understanding of the security state and risk posture of the information systems. They allow FedRAMP authorizing officials to make credible risk-based decisions regarding the continued operations of the information systems and initiate appropriate responses as needed when changes occur. The deliverable frequencies are to be considered standards. However, there will be instances, beyond the control of FedRAMP in which deliverables may be required on an ad hoc basis.

Routine Systems Change Control Process

The Change Control Process is instrumental in ensuring the integrity of the cloud computing environment. As the system owners—as well as other authorizing officials—approve changes, they are systematically documented. This documentation is a critical aspect of continuous monitoring because it establishes all of the requirements that led to the need for the change as well as the specific details of the implementation. To ensure that changes to the enterprise do not alter the security posture beyond the parameters set by the FedRAMP Joint Authorization Board (JAB), the key documents in the authorization package—which include the security plan, security assessment report, and plan of action and milestones—are updated and formally submitted to FedRAMP within 30 days of approved modification.

There are, however, changes that are considered to be routine. These changes can be standard maintenance, addition or deletion of users, the application of standard security patches, or other routine activities. While these changes individually may not have much effect on the overall security posture of the system, in aggregate they can create a formidable security issue. To combat this possibility, these routine changes should be documented as part of the CSP's standard change management process and accounted for via the CSP's internal continuous monitoring plan. Accordingly, these changes must be documented, at a minimum, within the current SSP of the system within 30 days of implementation.

Configuration Change Control Process (CCP)

Throughout the System Development Life Cycle (SDLC) system owners must be cognizant of changes to the system. Since systems routinely experience changes over time to accommodate new requirements, new technologies, or new risks, they must be routinely analyzed in respect to the security posture. Minor changes typically have little impact to the security posture of a system. These changes can be standard maintenance, adding or deleting users, applying standard security patches, or other routine activities. However, significant changes require an added level of attention and action. NIST defines significant change as, "A significant change is defined as a change that is likely to affect the security state of an information system." Changes such as installing a new operating system, port modification, new hardware platforms, or changes to the security controls should automatically trigger a re-authorization of the system via the FedRAMP process. Minor changes must be captured and documented in the SSP of the system within 30 days of implementation. This requirement should be part of the CSP's documented internal continuous monitoring plan. Once the SSP is updated, it must be submitted to FedRAMP, and a record of the change must be maintained internally. Major or significant changes may require re-authorization via the FedRAMP process. In order to facilitate a re-authorization, it is the responsibility of both the CSP and the sponsoring agency to notify FedRAMP of the need to make such a significant change. FedRAMP will assist with, and coordinate with all stakeholders, the necessary steps to ensure that the change is adequately documented, tested, and approved.

FISMA Reporting Requirements

FISMA established IT security reporting requirements. OMB, in conjunction with DHS, enforces these reporting requirements. FISMA reporting responsibilities must be clearly defined. FedRAMP will coordinate with CSPs and agencies to gather data associated with the cloud

service offering. Only data related to the documented system security boundary of the cloud service offering will be collected by FedRAMP and reported to OMB at the appropriate time and frequency. Agencies will maintain their reporting responsibilities for their internal systems that correspond to the inter-connection between the agency and the cloud service offering.

On-going Testing of Controls and Changes to Security Controls Process

System owners and administrators have long maintained the responsibility for patch and vulnerability management. However, it has been proven time and again that this responsibility often requires a heavy use of resources as well as a documented, repeatable process to be carried out consistently and adequately. This strain on resources and lack of processes has opened the door to many malicious entities through improper patching, significant lapse in time between patch availability and patch implementation, and other security oversights. Routine system scanning and reporting is a vital aspect of continuous monitoring and maintaining a robust cyber security posture.

Must Do

Vulnerability patching is critical. Proprietary operating system vendors (POSV) are constantly providing patches to mitigate vulnerabilities that are discovered. In fact, regularly scheduled monthly patches are published by many POSV to be applied to the appropriate operating system.

It is also the case that POSV will, from time to time, publish security patches that should be applied on systems as soon as possible due to the serious nature of the vulnerability. Systems running in virtual environment are not exempted from patching. In fact, not only are the operating systems running in a virtual environment to be patched routinely, but often the virtualization software itself is exposed to vulnerabilities and thus must be patched either via a vendor-based solution or other technical solution.

Open source operating systems require patch and vulnerability management as well. Due to the open nature of these operating systems there needs to be a reliable distribution point for system administrators to safely and securely obtain the required patches. These patches are available at the specific vendors' website.

Remember

Database platforms, web platforms and applications, and virtually all other software applications come with their own security issues. It is not only prudent, but also necessary, to stay abreast of all of the vulnerabilities that are represented by the IT infrastructure and applications that are in use. While vulnerability management is indeed a difficult and daunting task, there are proven tools available to assist the system owner and administrator in discovering the vulnerabilities in a timely fashion. These tools must be updated prior to being run. Updates are available at the corresponding vendors' website.

With these issues in mind, FedRAMP will require CSPs to provide the following:

- ☐ Monthly vulnerability scans of all servers. Tools used to perform the scan must be provided as well as the version number reflecting the latest update. A formal report of all vulnerabilities discovered, mitigated, or presentation of a mitigating strategy. This report should list the vulnerabilities by severity and name. Specificity is crucial to addressing the security posture of the system. All "High" level vulnerabilities must be mitigated within thirty days (30) days of discovery. "Moderate" level vulnerabilities must be mitigated within ninety (90) days of discovery. It is accepted that, at certain times, the application of certain security patches can cause negative effects on systems. In these situations, it is understood that compensating controls (workarounds) must be used to minimize system performance degradation while serving to mitigate the vulnerability. These workarounds must be submitted to FedRAMP and the sponsoring agency for acceptance. All reporting must reflect these activities.
- ☐ Quarterly FDCC and/or system configuration compliance scans, with a Security Content Automation Protocol (SCAP) validated tool, across the entire boundary, which verifies that all servers maintain compliance with the mandated FDCC and/or approved system configuration security settings.
- ☐ Weekly scans for malicious code. Internal scans must be performed with the appropriate updated toolset. Monthly reporting is required to be submitted to FedRAMP, where activity is summarized.
- ☐ All software operating systems and applications are required to be scanned by an appropriate tool to perform a thorough code review to discover malicious code. Mandatory reporting to FedRAMP must include tool used, tool configuration settings, scanning parameters, application scanned (name and version), and the name of the third party performing the scan. Initial report should be included with the SSP as part of the initial authorization package.
- ☐ Performance of the annual Self-Assessment in accordance with NIST guidelines. CSPs must perform a self-assessment annually, or whenever a significant change occurs. This is necessary if there is to be a continuous awareness of the risk and security posture of the system.
- ☐ Quarterly POA&M remediation reporting. CSPs must provide to FedRAMP a detailed matrix of POA&M activities using the supplied FedRAMP POA&M Template. This should include milestones met or milestones missed, resources required, and validation parameters.
- ☐ Active Incident Response capabilities allow for suspect systems to be isolated and inspected for any unapproved or otherwise malicious applications.

- ☐ Quarterly boundary-wide scans are required to be performed on the defined boundary IT system inventory to validate the proper HW and SW configurations as well as search and discover rogue systems attached to the infrastructure. A summary report, inclusive of a detailed network architecture drawing, must be provided to FedRAMP.
- ☐ Change Control Process meetings to determine and validate the necessity for suggested changes to HW/SW within the enterprise must be coordinated with FedRAMP to ensure that the JAB is aware of the changes being made to the system.

Incident Response

Remember

Computer security incident response has become an important component of information technology (IT) programs. Security-related threats have become not only more numerous and diverse, but also more damaging and disruptive. New types of security-related incidents emerge frequently. Preventative activities based on the results of risk assessments can lower the number of incidents, but not all incidents can be prevented. An incident response capability is therefore necessary for rapidly detecting incidents, minimizing loss and destruction, mitigating the weaknesses that were exploited, and restoring computing services. To that end, NIST SP 800-61 provides guidelines for development and initiation of an incident handling program, particularly for analyzing incident-related data and determining the appropriate response to each incident.

The guidelines can be followed independently of particular hardware platforms, operating systems, protocols, or applications. As part of the authorization process the system security plan will have documented all of the "IR" or Incident Response family of controls. One of these controls (IR-8) requires the development of an Incident Response plan that will cover the life cycle of incident response as documented in the NIST SP 800-61 guidelines. The plan should outline the resources and management support that is needed to effectively maintain and mature an incident response capability. The incident response plan should include these elements:

- ☐ Mission
- ☐ Strategies and goals
- ☐ Senior management approval
- ☐ Organizational approach to incident response
- ☐ How the incident response team will communicate with the rest of the organization
- ☐ Metrics for measuring the incident response capability
- ☐ Roadmap for maturing the incident response capability
- ☐ How the program fits into the overall organization.

Must Do

The organization's mission, strategies, and goals for incident response should help in determining the structure of its incident response capability. The incident response program structure should also be discussed within the plan. The response plan must address the possibility that incidents, including privacy breaches and classified spills, may impact the cloud and shared cloud customers. In any shared system, communication is the biggest key to success.

As part of the continuous monitoring of a system, responding to incidents will be a key element.

The FedRAMP concern and its role in continuous monitoring will be to focus on how a provider conducted the incident response and any after-incident actions. Incident response is a continually improving process.

One of the most important parts of incident response is also the most often omitted - learning and improving.

Each incident response team should evolve to reflect new threats, improved technology, and lessons learned. Many organizations have found that holding a "lessons learned" meeting with all involved parties after a major incident, and periodically after lesser incidents, is extremely helpful in improving security measures and the incident handling process itself. This meeting provides a chance to achieve closure with respect to an incident by reviewing what occurred, what was done to intervene, and how well intervention worked. The meeting should be held within several days of the end of the incident. Questions to be answered in the lessons learned meeting include:

- ☐ Exactly what happened, and at what times?
- ☐ How well did staff and management perform in dealing with the incident? Were the documented procedures followed? Were they adequate?
- ☐ What information was needed sooner?
- ☐ Were any steps or actions taken that might have inhibited the recovery?
- ☐ What would the staff and management do differently in a future occurrence?
- ☐ What corrective actions can prevent similar incidents in the future?
- ☐ What tools/resources are needed to detect, analyze, and mitigate future incidents?

Small incidents need limited post-incident analysis, with the exception of incidents performed through new attack methods that are of widespread concern and interest. After serious attacks have occurred, it is usually worthwhile to hold post-mortem meetings that cross team and organizational boundaries to provide a mechanism for information sharing. The primary consideration in holding such meetings is ensuring that the right people are involved. Not only is it important to invite people who have been involved in the incident that is being analyzed, but also wise to consider who should be invited for the purpose of facilitating future cooperation.

Independent Verification and Validation

Independent Verification and Validation (IV&V) is going to be an integral component to a successful implementation of FedRAMP. With this in mind, it must be noted that establishing and maintaining an internal expertise of FedRAMP policies, procedures, and processes is going to be required. This expertise will be tasked to perform various IV&V functions with CSPs, sponsoring agencies and commercial entities obtained by CSPs with absolute independence on behalf of

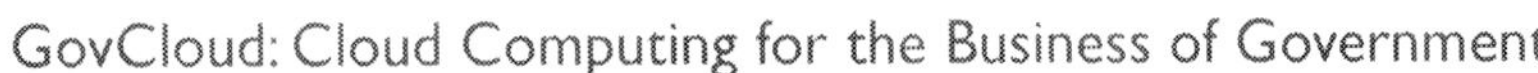

FedRAMP. FedRAMP IV&V will be on behalf of the JAB. As part of these efforts, FedRAMP will periodically perform audits (both scheduled and unscheduled) related strictly to the cloud computing service offering and the established system boundary. This will include, but not be limited to:

- ☐ Scheduled annual assessments of the system security documentation;
- ☐ Verification of testing procedures;
- ☐ Validation of testing tools and assessments;
- ☐ Validation of assessment methodologies employed by the CSP and independent assessors;
- ☐ Verification of the CSP continuous monitoring program; and
- ☐ Validation of CSP risk level determination criteria.

There are several methods that must be employed to accomplish these tasks. In accordance with the new FIMSA requirement, and as a matter of implementing industry best practices, FedRAMP IV&V will be performing penetration testing. This testing will be performed with strict adherence to the specific guidelines established by a mutually agreed upon "Rules of Engagement" agreement between FedRAMP IV&V and the target stakeholders. Unless otherwise stated in the agreement, all penetration testing will be passive in nature to avoid unintentional consequences. No attempts to exploit vulnerabilities will be allowed unless specified within the "Rules of Engagement" agreement.

FedRAMP Assessment and Authorization Process

The following figure depicts the high-level process for getting on the FedRAMP authorization request log. Once the Cloud Service Provider (CSP) system is officially on the FedRAMP authorization log, FedRAMP begins processing the cloud system for JAB authorization. The subsequent sections detail the steps involved in the FedRAMP Assessment and Authorization process.

Figure 13. FedRAMP Assessment Process

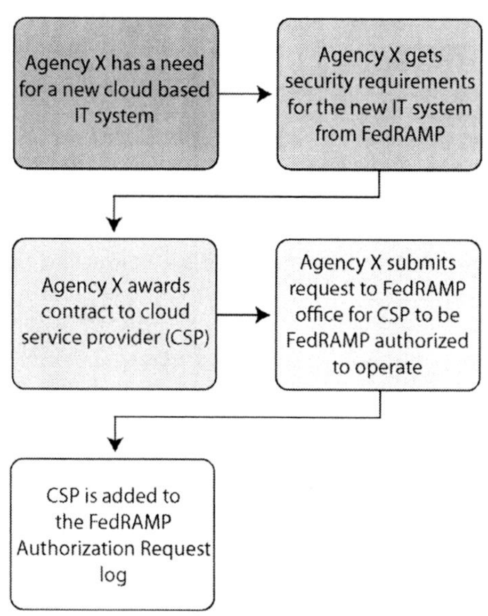

Figure 14. FedRAMP Authorization Process

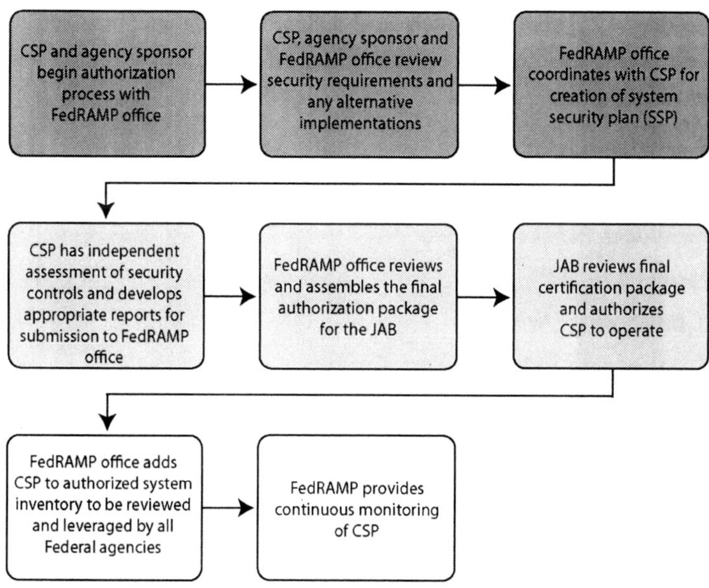

The FedRAMP Authorizing Officials (AO) must authorize, in writing, all cloud computing systems before they go into operational service for government interest.

A service provider's cloud computing systems must be authorized/reauthorized at least every three (3) years or whenever there is a significant change to the system's security posture in accordance with NIST SP 800-37 R1.

Tips

Authorization termination dates are influenced by FedRAMP policies that may establish maximum authorization periods. For example, if the maximum authorization period for an information system is three years, then the service provider establishes a continuous monitoring strategy for assessing a subset of the security controls employed within, and inherited by, the system during the authorization period. This strategy allows all security controls designated in the respective security plans to be assessed at least one time by the end of the three-year period.

This also includes any common controls deployed external to service provider cloud computing systems. If the security control assessments are conducted by qualified assessors with the required degree of independence based on policies, appropriate security standards and guidelines, and the needs of the FedRAMP authorizing officials, the assessment results can be cumulatively applied to the reauthorization, thus supporting the concept of ongoing authorization. FedRAMP policies regarding ongoing authorization and formal reauthorization, if/when required, are consistent with federal directives, regulations, and/or policies.

Required FedRAMP Artifacts

All service providers' CCS must complete and deliver the following artifacts as part of the authorization process. Templates for these artifacts can be found in FedRAMP templates as described in reference materials:

- ☐ Privacy Impact Assessment (PIA)
- ☐ FedRAMP Test Procedures and Results
- ☐ Security Assessment Report (SAR)
- ☐ System Security Plan (SSP)
- ☐ IT System Contingency Plan (CP)
- ☐ IT System Contingency Plan (CP) Test Results
- ☐ Plan of Action and Milestones (POA&M)
- ☐ Continuous Monitoring Plan (CMP)
- ☐ FedRAMP Control Tailoring Workbook
- ☐ Control Implementation Summary Table
- ☐ Results of Penetration Testing
- ☐ Software Code Review
- ☐ Interconnection Agreements/Service Level Agreements/Memorandum of Agreements

FedRAMP Guidance

The following is a list of NIST special publications, FIPS publications, OMB Memorandums, FedRAMP templates and other guidelines and documents associated with the seven steps of the FedRAMP process:

- ☐ **Step 1 - Categorize Cloud System:** (FIPS 199 / NIST Special Publications 800-30, 800-39, 800-59, 800-60.)
- ☐ **Step 2 – Select Security Controls:** (FIPS Publications 199, 200; NIST Special Publications 800-30, 800-53 R3, FedRAMP security control baseline)
- ☐ **Step 3 – Authorization Request:** (FedRAMP primary Authorization Request letter, FedRAMP secondary authorization request letter)
- ☐ **Step 4 - Implement Controls:** (FedRAMP control tailoring workbook; Center for Internet Security (CIS); United States Government Configuration Baseline (USGCB); FIPS Publication 200; NIST Special Publications 800-30, 800-53 R3, 800-53A R1)
- ☐ **Step 5 – Assess Controls:** (FedRAMP Test Procedures: Center for Internet Security (CIS); United States Government Configuration Baseline (USGCB); NIST Special Publication 800-53A R1)
- ☐ **Step 6 – Authorize Cloud System:** OMB Memorandum 02-01; NIST Special Publications 800- 30, 800-53A R1)
- ☐ **Step 7 – Continuous Monitoring:** FedRAMP Test Procedures; NIST Special Publications 800-30, 800-53A R1, 800-37 R1

How Can I Buy A Cloud Service?

Business Apps

Your agency or service is complex and requires state-of-the-art software to get business done. GSA's Cloud Business Apps provide the solution. Business Apps is geared towards the enterprise with cloud software solutions such as analytical, business processes, CRM, tracking and monitoring tools, business intelligence, and more.

Cloud IT Services

You've got servers, storage, development, testing, and production teams - and on it goes. GSA's Cloud IT Services offer solutions to reduce cost and implement projects faster.

Productivity Apps

GSA Cloud productivity software and applications are geared towards government employees who need software to perform daily tasks such as word processing and spreadsheets but also includes brainstorming, collaboration, document management, project management and more.

Social Media Apps

Social media apps make it easier to create and distribute content, discuss the things we care about, and help people get their jobs done. Social media includes various online technology tools that enable people to communicate easily and share information. Social media includes text, audio, video, images, podcasts, and other multimedia communications.

Building a Cloud

Traditional federal system integrators can also help agencies identify and evaluate cloud computing opportunities. When contemplating this option, agencies should use the Government Cloud Computing Framework as a reference for discussing implementation designs. In order to ensure interoperability and portability, industry accepted open standards should be used to the maximum extent possible.

For more information, go to :https://apps.gov/cloud/advantage/main/start_page.do

Figure 15. Government Cloud Computing Framework

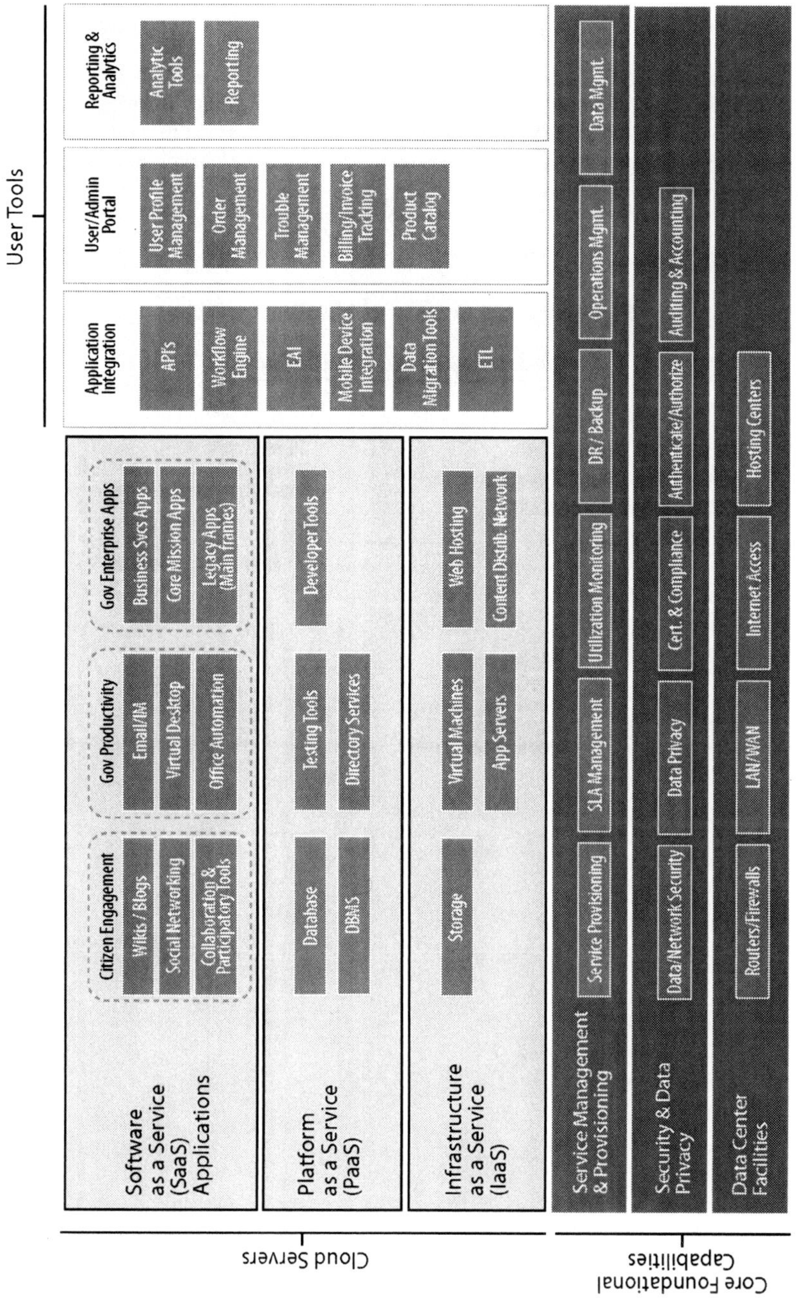

User Tools

- Application Integration
  - APIs
  - Workflow Engine
  - EAI
  - Mobile Device Integration
  - Data Migration Tools
  - ETL
- User/Admin Portal
  - User Profile Management
  - Order Management
  - Trouble Management
  - Billing/Invoice Tracking
  - Product Catalog
- Reporting & Analytics
  - Analytic Tools
  - Reporting

Cloud Servers

- Software as a Service (SaaS) Applications
  - Citizen Engagement
    - Wikis / Blogs
    - Social Networking
    - Collaboration & Participatory Tools
  - Gov Productivity
    - Email/IM
    - Virtual Desktop
    - Office Automation
  - Gov Enterprise Apps
    - Business Svcs Apps
    - Core Mission Apps
    - Legacy Apps (Main frames)
- Platform as a Service (PaaS)
  - Database
  - DBMS
  - Testing Tools
  - Directory Services
  - Developer Tools
- Infrastructure as a Service (IaaS)
  - Storage
  - Virtual Machines
  - App Servers
  - Web Hosting
  - Content Distrib. Network

Core Foundational Capabilities

- Service Management & Provisioning
  - Service Provisioning
  - SLA Management
  - Utilization Monitoring
  - DR / Backup
  - Operations Mgmt
  - Data Mgmt
- Security & Data Privacy
  - Data/Network Security
  - Data Privacy
  - Cert. & Compliance
  - Authenticate/Authorize
  - Auditing & Accounting
- Data Center Facilities
  - Routers/Firewalls
  - LAN/WAN
  - Internet Access
  - Hosting Centers

Cloud Computing Standards

Cloud Standards Coordination

Remember

Although it is still in early stages of development, there are many industry groups working to develop and coordinate cloud standards. This includes the development of a Cloud Landscape to overview the various efforts and introduce terms and definitions that allow each standard to be described in common language, and an entry for each standard categorized by organization.

Cloud Standards Overview

Figure 16. Taxonomy of cloud interfaces

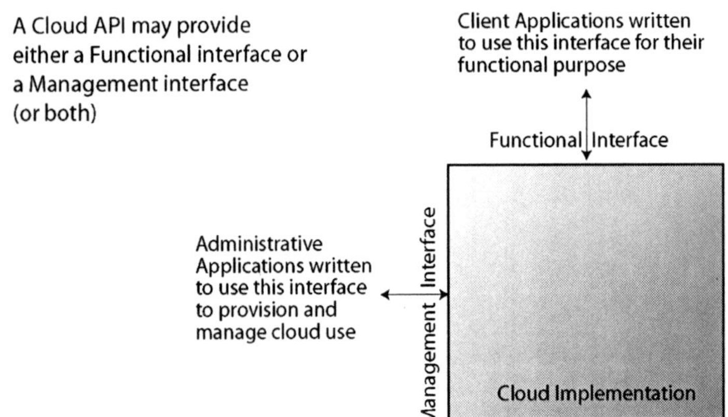

Figure 17. The stack

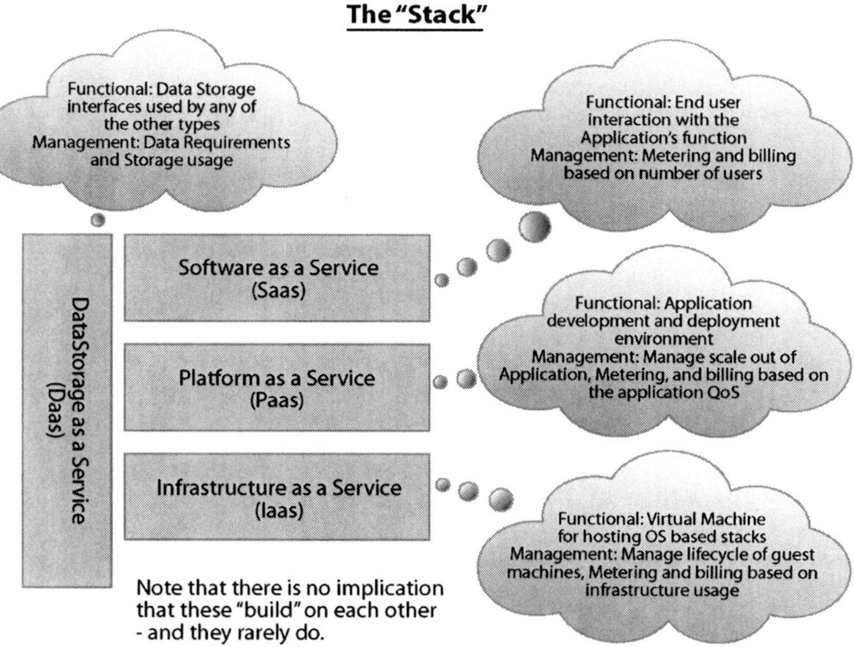

Figure 18. Cloud management

Cloud Management

Cloud Management has multiple aspects that can be standardized for interoperability

For a given type of cloud (e.g. Iaas) the standards could be split across different SDOs

Some Examples:
- Provisioning
- Metering and Billing
- Security
- Privacy
- Quality of Service (QoS)
- Identity

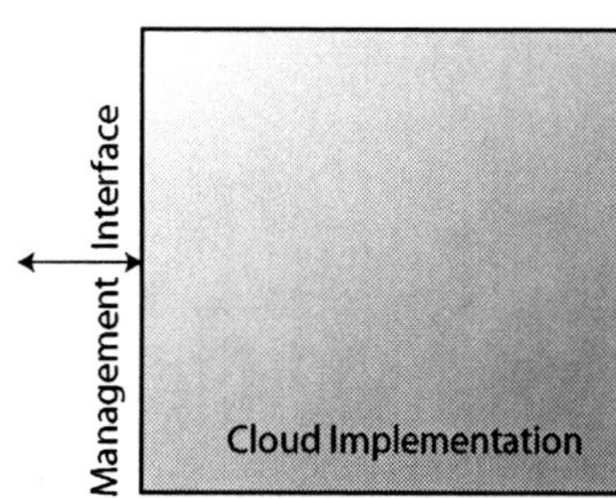

Some possible standards:

- ☐ Federated security (e.g. identity) across clouds
- ☐ Metadata and data exchanges across clouds
- ☐ Standards for moving applications between cloud platforms
- ☐ Standards for describing resource/performance capabilities and requirements
- ☐ Standardized outputs for monitoring, auditing, billing, reports, and notification for cloud applications and services
- ☐ Common representations (abstracts, APIs, protocols) for interfacing to cloud resources
- ☐ Cloud-independent representation for policies and governance
- ☐ Portable tools for developing, deploying, and managing cloud applications and services
- ☐ Orchestration and middleware tools for creating composite applications across clouds
- ☐ Standards for machine-readable Service Level Agreements (SLAs)

Standards and Test Bed Groups

- ☐ Cloud Security Alliance (CSA)
- ☐ Distributed Management Task Force (DMTF)
- ☐ Storage Networking Industry Association (SNIA)
- ☐ Open Grid Forum (OGF)
- ☐ Open Cloud Consortium (OCC)
- ☐ Organization for the Advancement of Structured Information Standards (OASIS)
- ☐ TM Forum
- ☐ Internet Engineering Task Force (IETF)
- ☐ International Telecommunications Union (ITU)
- ☐ European Telecommunications Standards Institute (ETSI)
- ☐ Object Management Group (OMG)

Cloud Standards Positioning

One interesting proposed initiative is sharing a general "cloud computing standardization positioning" site in which cloud standardization initiatives can be uploaded and reviewed.

STEP FIVE. HOW TO MAKE SURE IT'S WORKING

Key Performance Indicators

Cloud computing introduces an expanded context for service-oriented business and IT. Developing ROI models that show how cloud computing adoption can benefit both business and IT consumers and providers involves examining the key technology features and business operating model changes.

This section gives an overview of ROI models to support cloud computing assessments and business cases in two aspects:

- ☐ Key Performance Indicator ratios that target cloud computing adoption, comparing specific metrics of traditional IT with cloud computing solutions. These have been classified as cost, time, quality, and profitability indicators relating to cloud computing characteristics.
- ☐ Key Return on Investment savings models that demonstrate cost, time, quality, compliance, revenue, and profitability improvement by comparing traditional IT with cloud computing solutions.

Cloud Computing ROI Models and KPI

Key Performance Indicators (KPI) - A set of quantifiable measures that a company or industry uses to gauge or compare performance in terms of meeting their strategic and operational goals. KPIs vary between companies and industries, depending on their priorities or performance criteria. Also referred to as "key success indicators (KSI)."

Return on Investment (ROI) - A performance measure used to evaluate the efficiency of an investment or to compare the efficiency of a number of different investments. To calculate ROI, the benefit (return) of an investment is divided by the cost of the investment; the result is expressed as a percentage or a ratio. The return on investment formula:

Equation 1. Return on Investment

$$\text{ROI} = \frac{(\text{Gain from Investment - Cost of Investment})}{\text{Cost of Investment}}$$

The overview of cloud computing ROI models considers both indicators and ROI viewpoints.

Figure 19. Cloud Computing ROI models & KPIs

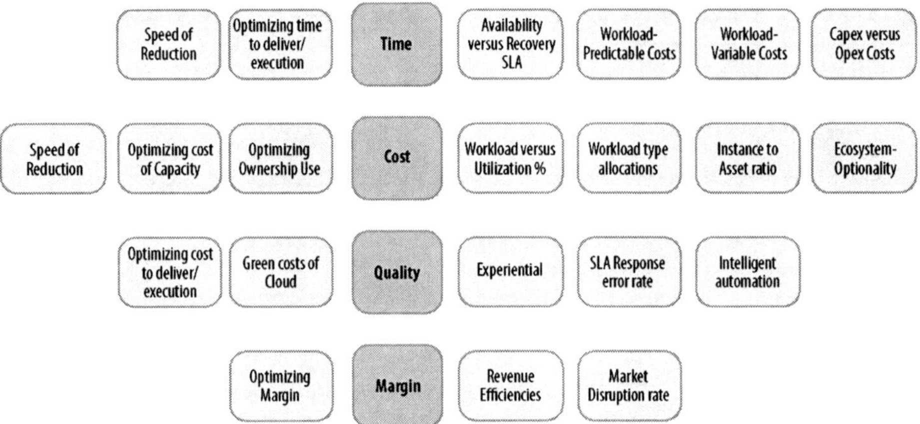

Cloud ROI Cost Indicator Ratios

Figure 20. Cost Indicator Ratios

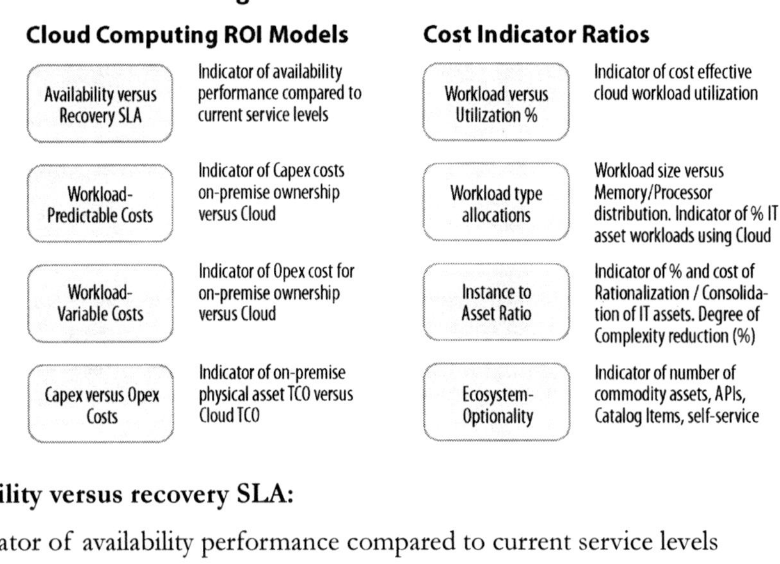

Availability versus recovery SLA:

☐ Indicator of availability performance compared to current service levels

Workload – predictable costs:

☐ Indicator of CAPEX cost on-premise ownership versus cloud

Workload – variable costs:

☐ Indicator of OPEX cost for on-premise ownership versus cloud; indicator of burst cost

CAPEX versus OPEX costs:

☐ Indicator of on-premise physical asset TCO versus cloud TCO

Workload versus utilization %:

☐ Indicator of cost-effective cloud workload utilization

Workload type allocations:

☐ Workload size versus memory/processor distribution; indicator of % IT asset workloads using cloud

Instance to asset ratio:

☐ Indicator of % and cost of rationalization/consolidation of IT assets; degree of complexity reduction

Ecosystem – optionality:

☐ Indicator of number of commodity assets, APIs, catalog items, self service

Cloud ROI Time Indicator Ratios

Figure 21. Time Indicator Ratios

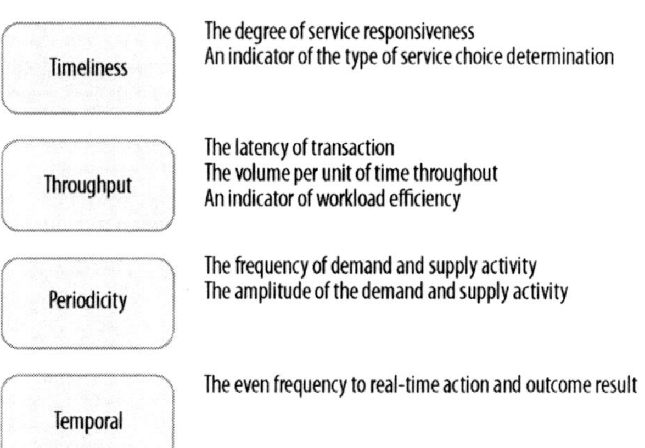

Timeliness:

☐ The degree of service responsiveness

☐ An indicator of the type of service choice determination

Throughput:

☐ The latency of transactions

☐ The volume per unit of time throughput

☐ An indicator of the workload efficiency

Periodicity:

☐ The frequency of demand and supply activity

☐ The amplitude of the demand and supply activity

Temporal:

☐ The event frequency to real-time action and outcome result

Cloud ROI Quality Indicator Ratios

Figure 22. Profitability Indicator Ratios

Cloud Computing ROI Models - Profitability Indicator Ratios

Revenue Efficiencies	The ability to generate margin increase per revenue Rate of annuity improvement
Market Disruption Rate	Rate of Revenue growth Rate of New Product market acquisition

Revenue efficiencies:

☐ Ability to generate margin increase/budget efficiency per margin

☐ Rate of annuity revenue

Market disruption rate:

☐ Rate of revenue growth

☐ Rate of new market acquisition

Cloud ROI Savings Models

Figure 23. Cloud Computing ROI Savings Models

Cloud Computing ROI - Savings Models

Time	Cost	Quality	Profitability
Speed of Reduction	Speed of Reduction	Green costs of Cloud	Optimizing Margin
Rate of change of TCO reduction by Cloud adoption	Rate of change of TCO reduction by Cloud adoption	Green sustainability	Increase in Revenue/ Profit margin from Cloud adoption
Optimizing time to deliver / execution	Optimizing cost of Capacity	Optimizing time to deliver / execution	
Increase in Provisioning speed Speed of multi-sourcing	Aligning cost with usage. Capex to Opex Utilization pay-as-you-go savings from Cloud adoption	Reduced supply chain costs Flexibility / Choice	
	Optimizing Ownership Use		
	Portfolio TCO License cost reduction from Cloud adoption Open Source Adoption SOA Resue Adoption		

Speed of time reduction:

☐ Compression of time reduction by cloud adoption

☐ Rate of change of TCO reduction by cloud adoption

Optimizing time to deliver/execution:

☐ Increase in provisioning speed

☐ Speed of multi-sourcing

Speed of cost reduction:

☐ Compression of cost reduction by cloud adoption

☐ Rate of change of TCO reduction by cloud adoption

Optimizing cost of capacity:

☐ Aligning cost with usage, CAPEX to OPEX utilization pay-as-you-go savings from cloud adoption

☐ Elastic scaling cost improvements

Optimizing ownership use:

☐ Portfolio TCO , license cost reduction from cloud adoption

☐ Open source adoption

☐ SOA re-use adoption

Green costs of Cloud:

☐ Green sustainability

Optimizing time to deliver/execution:

☐ Increase in provisioning speed

☐ Reduced supply chain costs

☐ Speed of multi-sourcing

☐ Flexibility/choice

Optimizing margin:

☐ Increase in revenue/profit margin from cloud adoption

Cloud Computing Quick Reference

Cloud Security Alliance

The Cloud Security Alliance was created to promote the use of best practices for providing security assurance within cloud computing, and provide education on the uses of cloud computing to help secure all other forms of computing.

Security Guidance for Critical Areas of Focus in Cloud Computing

Covers key issues and provides advice for both cloud computing customers and providers within 15 strategic domains.

Cloud Controls Matrix

The Cloud Security Alliance Controls Matrix (CM) is specifically designed to provide fundamental security principles to guide cloud vendors and to assist prospective cloud customers in assessing the overall security risk of a cloud provider.

Top Threats to Cloud Computing

The purpose of Top Threats to Cloud Computing is to provide needed context to assist organizations in making educated risk management decisions regarding their cloud adoption strategies.

Remember

CloudAudit

The goal of CloudAudit is to provide a common interface and namespace that allows cloud computing providers to automate the Audit, Assertion, Assessment, and Assurance (A6) of their infrastructure (IaaS), platform (PaaS), and application (SaaS) environments, and allow authorized consumers of their services to do likewise via an open, extensible, and secure interface and methodology.

Distributed Management Task Force (DMTF)

Open Virtualization Format (OVF)

- ☐ DSP0243 Open Virtualization Format (OVF) V1.1.0
- ☐ OVF has been designated as ANSI INCITS 469 2010

This specification describes an open, secure, portable, efficient and extensible format for the packaging and distribution of software to be run in virtual machines.

Open Cloud Standards Incubator

DMTF's Open Cloud Standards Incubator focused on standardizing interactions between cloud environments by developing cloud management use cases, architectures and interactions. This work was completed in July 2010. The work has now transitioned to the Cloud Management Working Group.

Interoperable Clouds White Paper

- ☐ DSP-IS0101 Cloud Interoperability White Paper V1.0.0

This white paper describes a snapshot of the work being done in the DMTF Open Cloud Standards Incubator, including use cases and reference architecture as they relate to the interfaces between a cloud service provider and a cloud service consumer.

Architecture for Managing Clouds White Paper

☐ DSP-IS0102 Architecture for Managing Clouds White Paper V1.0.0

This white paper is one of two Phase 2 deliverables from the DMTF Cloud Incubator and describes the reference architecture as it relates to the interfaces between a cloud service provider and a cloud service consumer. The goal of the Incubator is to define a set of architectural semantics that unify the interoperable management of enterprise and cloud computing.

Use Cases and Interactions for Managing Clouds White Paper

☐ DSP-IS0103 Use Cases and Interactions for Managing Clouds White Paper V1.0.0

This document is one of two documents that together describe how standardized interfaces and data formats can be used to manage clouds. This document focuses on use cases, interactions, and data formats.

Cloud Management Working Group (CMWG)

The CMWG will develop a set of prescriptive specifications that deliver architectural semantics as well as implementation details to achieve interoperable management of clouds between service requestors/developers and providers. This WG will propose a resource model that at minimum captures the key artifacts identified in the Use Cases and Interactions for Managing Clouds document produced by the Open Cloud Incubator.

The European Telecommunications Standards Institute (ETSI)

TC CLOUD

The goal of ETSI TC CLOUD (previously TC GRID) is to address issues associated with the convergence between IT (Information Technology) and Telecommunications. The focus is on scenarios where connectivity goes beyond the local network. This includes not only grid computing but also the emerging commercial trend towards cloud computing which places particular emphasis on ubiquitous network access to scalable computing and storage resources.

Since TC CLOUD has particular interest in interoperable solutions in situations which involve contributions from both the IT and Telecom industries, the emphasis is on the Infrastructure as a Service (IaaS) delivery model. TC GRID focuses on interoperable applications and services based on global standards and the validation tools to support these standards. Evolution towards a coherent and consistent general purpose infrastructure is envisaged. This will support networked IT applications in business, public sector, academic, and consumer environments.

ETSI Terms and Diagrams

National Institute of Standards and Technology (NIST)

NIST Working Definition of Cloud Computing

NIST is posting its working definition of cloud computing that serves as a foundation for its upcoming publication on the topic (available above). Computer scientists at NIST developed this definition in collaboration with industry and government. It was developed as the foundation for a NIST special publication that will cover cloud architectures, security, and deployment strategies for the federal government.

Standards Acceleration to Jumpstart Adoption of Cloud Computing (SAJACC)

Three complementary activities all performed in collaboration with other agencies and standards development organizations:

1. NIST inserts existing standards and de-facto interfaces as specifications.
   - NIST identifies and validates specifications using use cases.
2. Organizations contribute open specifications.
   - NIST receives and coordinates the prioritization of specifications, and validates using use cases.
3. NIST identifies gaps in cloud standards (and specifications) and publishes the gaps on the portal: produces opportunity for outside organizations to fill them.

Cloud Computing Use Cases

A set of twenty-five use cases that seek to express selected portability, interoperability, and security concerns that cloud users may have.

Open Grid Forum (OGF)

Open Cloud Computing Interface (OCCI) Working Group

The purpose of this group is the creation of a practical solution to interface with cloud infrastructures exposed as a service (IaaS). It will focus on a solution which covers the provisioning, monitoring and definition of cloud infrastructure services. Overlapping work and efforts will be contributed and synchronized with other groups.

Open Cloud Computing Interface Specification

Open Cloud Computing Interface Terms and Diagrams

OGF and SNIA have collaborated on a Cloud Storage for Cloud Computing white paper.

Object Management Group (OMG)

OMG's focus is always on modeling, and the first specific cloud-related specification efforts have only just begun, focusing on modeling deployment of applications and services on clouds for portability, interoperability and reuse.

Open Cloud Consortium (OCC)

- ☐ Supports the development of standards for cloud computing and frameworks for interoperating between clouds;
- ☐ Develops benchmarks for cloud computing; and
- ☐ Supports reference implementations for cloud computing, preferably open source reference implementations.

The OCC has a particular focus in large data clouds. It has developed the MalStone Benchmark for large data clouds and is working on a reference model for large data clouds.

Organization for the Advancement of Structured Information Standards (OASIS)

OASIS drives the development, convergence, and adoption of open standards for the global information society. The source of many of the foundational standards in use today, OASIS sees cloud computing as a natural extension of SOA and network management models. The OASIS technical agenda is set by members, many of whom are deeply committed to building cloud models, profiles, and extensions on existing standards, including:

- ☐ Security, access and identity policy standards -- e.g., OASIS SAML, XACML, SPML, WS-SecurityPolicy, WS-Trust, WS-Federation, KMIP, and ORMS.
- ☐ Content, format control and data import/export standards -- e.g., OASIS ODF, DITA, CMIS, and SDD.
- ☐ Registry, repository and directory standards -- e.g., OASIS ebXML and UDDI.
- ☐ SOA methods and models, network management, service quality and interoperability -- e.g., OASIS SCA, SDO, SOA-RM, and BPEL.

OASIS Cloud-Specific or Extended Technical Committees

- ☐ OASIS Identity in the Cloud (IDCloud) TC

The OASIS IDCloud TC works to address the serious security challenges posed by identity management in cloud computing. The TC identifies gaps in existing identity management standards and investigates the need for profiles to achieve interoperability within current standards. It performs risk and threat analyses on collected use cases and produces guidelines for mitigating vulnerabilities.

OASIS Symptoms Automation Framework (SAF) TC

Cloud computing, in particular, exacerbates the separation between consumer-based business requirements and provider-supplied IT responses. The SAF facilitates knowledge sharing across these domains, allowing consumer and provider to work cooperatively together to ensure adequate capacity, maximize quality of service, and reduce cost.

Storage Networking Industry Association (SNIA)

SNIA Cloud TWG

The SNIA has created the Cloud Storage Technical Work Group for the purpose of developing SNIA Architecture related to system implementations of Cloud Storage technology. The Cloud Storage TWG:

- ☐ Acts as the primary technical entity for the SNIA to identify, develop, and coordinate systems standards for cloud atorage.
- ☐ Produces a comprehensive set of specifications and drives consistency of interface standards and messages across the various cloud storage-related efforts.
- ☐ Documents system-level requirements and shares these with other cloud storage standards organizations under the guidance of the SNIA Technical Council and in cooperation with the SNIA Strategic Alliances Committee.

SNIA Cloud Data Management Interface (CDMI)

The CDMI specification is now a SNIA Architecture standard and will be submitted to the INCITS organization for ratification as an ANSI and ISO standard as well.

SNIA CDMI Reference Implementation

The first working draft release of the Reference Implementation of CDMI is now available for download at Snia.org/cloud.

SNIA Terms and Diagrams

SNIA and OGF have collaborated on a Cloud Storage for Cloud Computing white paper. A demo of this architecture has been implemented and shown several times. More information can be found at the Cloud Demo Google Group.

Cloud Data Management Interface (CDMI) now has a working draft reference implementation available. Download and implement: Snia.org/cloud

The Open Group

Cloud Work Group

The Cloud Work Group exists to create a common understanding among buyers and suppliers of how enterprises of all sizes and scales of operation can include cloud computing technology in a safe and secure way in their architectures to realize its significant cost, scalability, and agility benefits. It includes some of the industry's leading cloud providers and end-user organizations, collaborating on standard models and frameworks aimed at eliminating vendor lock-in for enterprises looking to benefit from cloud products and services.

The Open Group Cloud Work Group has established several projects to enhance business understanding, analysis and uptake of cloud computing technologies, including:

- ☐ Cloud Business Use Cases
- ☐ Cloud Business Artifacts
- ☐ Cloud Computing Architecture
- ☐ Service Oriented Cloud Computing Infrastructure
- ☐ Security in the Cloud

Going forward, the group plans to provide a set of tools and templates to support business decisions on cloud computing, including:

- ☐ Cloud Business Use Case Template
- ☐ Cloud Taxonomy for Buyers
- ☐ Cloud Taxonomy for Sellers
- ☐ CC Financial and ROI Templates
- ☐ CC Business Adoption Strategies
- ☐ Cloud definitions for business

Cloud Computing Business Scenario

The Business Scenario technique of TOGAF™ can be used to gather and represent customer requirements in order for the supply-side to better understand real needs of the customer-side. The purpose of the Business Scenario is to gather customer views on the motivations for, and key requirements of, the use of cloud computing technologies.

- ☐ Building Return on Investment from Cloud Computing

This white paper presents the initial conclusions from The Open Group on how to build and measure Return on Investment (ROI) from Cloud Computing. It was produced by the Cloud Business Artifacts (CBA) project of The Open Group Cloud Computing Work Group.

Association for Retail Technology Standards (ARTS)

ARTS has recently announced their Cloud Computing White Paper V1.0. This Cloud Computing for Retail white paper offers unbiased guidance for achieving maximum results from this relatively new technology. Version 1.0 represents a significant update to the draft version released in October 2009, specifically providing more examples of cloud computing in retail, as well as additional information on the relationship to Service Oriented Architecture (SOA) and constructing a Private Cloud.

TM Forum

The TM Forum is a not-for-profit Global Industry Association with over 750 members in 195 countries, including the world's largest service providers, enterprise customers, hardware and software companies, SIs, and consulting companies. TM Forum is the leading industry association focused on enabling best-in-class IT for service providers in the communications, media, and cloud service markets. The Forum provides business-critical industry standards and expertise to enable the creation, delivery, and monetization of digital services.

TM Forum brings together communications, cloud, technology and media companies, providing an innovative, industry-leading approach to collaborative R&D, along with wide range of support services including benchmarking, training, and certification. The Forum produces the renowned international Management World conference series, as well as thought-leading industry research and publications.

Cloud Services Initiative

Cloud services represent a significant evolution in the use and provision of digital information services for business effectiveness. Yet as buyers start to look at using these services, it is clear there are a number of barriers to adoption.

The primary objective of TM Forum's Cloud Services Initiative is to help the industry overcome these barriers and assist in the growth of a vibrant commercial marketplace for cloud-based services. The centerpiece of this initiative is an ecosystem of major buyers and sellers who will collaborate to define a range of common approaches, processes, metrics, and other key service enablers.

TM Forum's Cloud Services Initiative Vision

The TM Forum's Cloud Services Initiative aims to stimulate growth of a vibrant and open marketplace for cloud services by bringing together the entire ecosystem of enterprise customers, cloud service providers, and technology suppliers to remove barriers to adoption based on industry standards.

The Cloud Services Initiative delivers:

- ☐ An ecosystem of enterprise customers, cloud service providers, and technology suppliers that enable the commercialization of this major business opportunity
- ☐ Business guidance including benchmarks and service quality metrics
- ☐ Technical agreements – many in collaboration with other industry groups

Barriers to Success

The benefits of cloud computing, and its potential application in so many industry verticals, is creating a dramatic shift in the marketplace with many major suppliers of software and hardware reshaping around a cloud operating model. A lack of requirements from the buy-side of the equation will only lead to a repeat the past—opaque pricing, inconsistent offerings, and a lack of alignment with buyer needs.

By organizing and clearly articulating buyers requirements, and then bringing buyers and sellers together to agree required standards, the Enterprise Cloud Leadership Council (ECLC) seeks to accelerate the effective adoption of cloud computing on a global scale. The ECLC is shaping the future of enterprise IT by consolidating the requirements of the world's largest enterprises within the TM Forum and sits at the heart of the Cloud Services Initiative Program.

Enterprise Cloud Leadership Council Goals (ECLC)

- ☐ To foster an effective and efficient marketplace for cloud computing infrastructure and services across all industry verticals and global geographies;
- ☐ Accelerate standardization and commoditization of cloud services, and identifying common commodity processes best consumed as a service;
- ☐ Solicit definition for standardized core and industry-specific SKUs for cloud services;
- ☐ Achieve transparency of cost, service levels, and reporting across the ecosystem;
- ☐ Enable benchmarking of services across service providers and geographies;
- ☐ Enable vendor measurement against normalized and agreed service level metrics

Future Collaborative Programs

- ☐ Defining Service Level Agreements for cloud services
- ☐ Database-as-a-Service (DBaaS) reference architecture
- ☐ Cloud API requirements
- ☐ Business Process and Information Frameworks for Cloud
- ☐ Secure Virtual Private Cloud reference architecture
- ☐ Standard service definitions/SKUs (taxonomy of services)

- ☐ Cloud SDO liaisons
- ☐ eTOM and ITIL: how to combine them in a cloud context
- ☐ Cloud service provider benchmarking and metrics
- ☐ Billing engines; client billing and partner revenue sharing for cloud services
- ☐ Common definition of commercial terms (business contract language)

TM Forum's Frameworx

TM Forum's Frameworx Integrated Business Architecture provides an industry agreed, service-oriented approach for rationalizing operational IT, processes, and systems that enables service providers to significantly reduce their operational costs and improve business agility.

Service-oriented approaches encourage business agility through reuse, which is essential in today's market where service providers need to deliver new services rapidly and increase revenues in the face of changing value chains and technologies.

Frameworx uses standard, reusable, generic blocks—Platforms and Business Services—that can be assembled in unique ways to gain the advantages of standardization while still allowing customization where necessary.

Frameworx defines the mechanism by which the Forum's existing NGOSS standard framework components are integrated into a comprehensive enterprise IT and process architecture that also embraces major IT industry standards such as ITIL and TOGAF. Its components are:

- ☐ Business Process Framework (eTOM) is the industry's common process architecture for both business and functional processes
- ☐ Information Framework (SID) provides a common reference model for enterprise information that service providers, software providers, and integrators use to describe management information
- ☐ Application Framework (TAM) provides a common language between service providers and their suppliers to describe systems and their functions, as well as a common way of grouping them
- ☐ Integration Framework provides a service-oriented integration approach with standardized interfaces and support tools

Additional tools

For more information go to: Nist.gov/itl/cloud/upload/RefArch-CCW-II.pdf

Figure 24. Additional Cloud Evaluation Tools

1. Cloud Application Suitability Model (CASM)

2. Wheel of Security

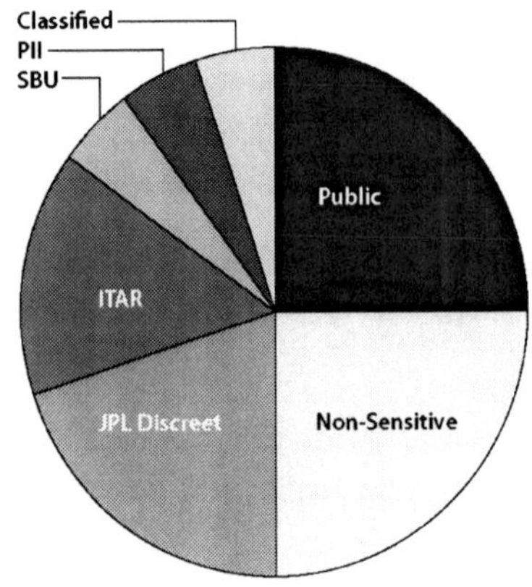

Public and Non-Sensitive data can be accessed in the Cloud today

3. Cloud Readiness Levels (CRL) (Institution, Apps, Dev)

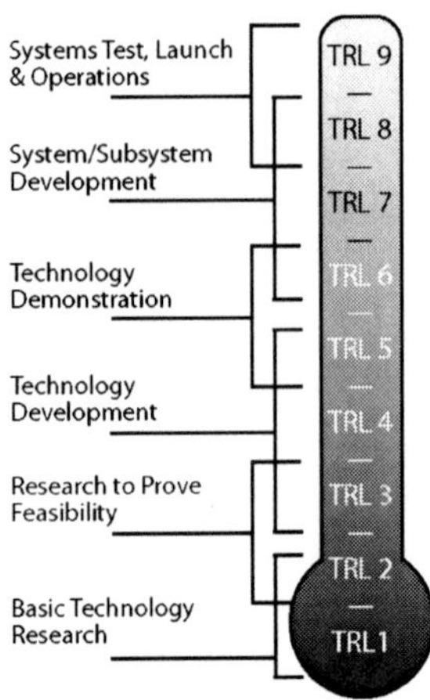

NASA Technology Readiness Level

Federal Cloud Computing Case Studies

Introduction

Remember

Cloud computing provides tremendous opportunities for the public sector to improve the delivery of services to the American people, reduce the cost of government operations, make more effective use of taxpayer dollars, and lower energy consumption. While the public sector is just at the beginning of the journey to cloud computing, we are already seeing innovative examples at all levels of government.

For example, on April 26, 2010, Recovery.gov became the first government-wide system to migrate to a cloud-based environment. With the cost savings gained from using a cloud computing infrastructure, the Recovery Board plans to redirect more than $1 million in computer equipment and software to its accountability mission to help identify fraud, waste, and abuse. The City of Los Angeles is anticipating savings of $5.5 million over five years as a result of moving e-mail and productivity tools to the cloud for over 34,000 city employees, and the State of Wisconsin's Department of Natural Resources is increasing collaboration through a hosted online meeting space that supports conference calls, interactive meetings, and information sharing.

These are a handful of illustrative examples that are part of a larger movement to leverage cloud computing across the public sector.

Use Cases

The following case studies provide recent examples of how federal agencies are using cloud computing technologies.

- ☐ Department of Defense (United States Army) - Army Experience Center
- ☐ Department of Defense (Defense Information Systems Agency) - Rapid Access Computing Environment
- ☐ Department of Defense (Defense Information Systems Agency) - Forge.mil
- ☐ Department of Defense (United States Air Force) - Personnel Services Delivery Transformation
- ☐ Department of Energy (Lawrence Berkeley National Labs) - Cloud Computing Migration
- ☐ Department of Health and Human Services - Supporting Electronic Health Records
- ☐ Department of the Interior - Agency-wide E-mail
- ☐ General Services Administration (Office of Citizen Services) - USA.gov
- ☐ General Services Administration - Agency-wide E-mail
- ☐ National Aeronautics and Space Administration (Ames Research Center) - World-Wide Telescope
- ☐ National Aeronautics and Space Administration (Jet Propulsion Laboratory) - Be A Martian
- ☐ National Aeronautics and Space Administration - Enterprise Data Center Strategy
- ☐ Social Security Administration - Online Answers Knowledgebase
- ☐ Federal Labor Relations Authority - Case Management System
- ☐ Recovery Accountability and Transparency Board - Recovery.gov Cloud Computing Migration
- ☐ Securities and Exchange Commission - Investor Advocacy System

Department of Defense

Project: Army Experience Center (United States Army)

The Army Experience Center (AEC), located in Philadelphia, PA, is an Army pilot program designed to explore new technologies and techniques that the Army can leverage to improve the efficiency and effectiveness of its marketing and recruiting operations. The AEC uses touch screen career exploration kiosks, state-of-the-art presentation facilities, community events, virtual reality simulators, and social networking to help potential recruits learn about the Army and make

informed decisions about enlisting. The Army required a customer relationship management system that would track personal and electronic engagements with prospects and would help recruiting staff manage the recruiting process.

Army's legacy proprietary data system, the Army Recruiting Information Support System (ARISS), was over 10 years old. Despite regular upgrades over the years, it was infeasible to modify ARISS to meet the AEC's requirements; including integration with social networking and other Web 2.0 applications, real-time data access from multiple platforms including hand-held devices, ability to track AEC visitor and engagement data, and integration of marketing and recruiting data. Initial bids from traditional IT vendors to provide required functionality ranged from $500,000 to over $1 million.

Instead, the Army chose a customized version of the cloud-based Customer Relationship Management tool offered by Salesforce.com as its pilot solution to manage recruiting efforts at the Army Experience Center. The Army is piloting this cloud-based solution at an annual cost of $54,000. With the new system, the Army is able to track recruits as they participate in multiple simulations at the Army Experience Center. The solution integrates directly with e-mail and Facebook, allowing recruiters to connect with participants more dynamically after they leave the Army Experience Center. By using Salesforce.com's mobile solution, Army recruiters can access recruit information from anywhere.

The Army is currently in the second year of a two-year pilot of the customized Salesforce.com application. Using the cloud-based solution, the Army was able to have fewer recruiters handle the same workload as the five traditional recruiting centers the Army Experience Center replaced. The cloud application has resulted in faster application upgrades, dramatically reduced hardware and IT staff costs, and significantly increased staff productivity.

Project: Rapid Access Computing Environment (Defense Information Systems Agency)

The Defense Information Systems Agency (DISA) provides IT support to the Department of Defense (DoD). DISA began leveraging cloud computing in 2008 by creating its own secure private cloud, the Rapid Access Computing Environment (RACE).

RACE, which uses virtual server technology to provide on-demand server space for development teams, aims to be more secure and stable than a traditional public cloud.

RACE consists of many virtual servers inside a single physical server. By using virtualization technologies, DISA has divided the costs of provisioning and operating a single physical server among the users of the various virtual servers. This system passes cost savings on to individual teams. Within this virtual environment, users can use a self-service portal to provision computing resources in 50 GB increments with the guarantee that the environment will be secure to DoD standards. At DoD, a dedicated server environment used to take three to six weeks to provision due to lengthy procurement processes. However, RACE is able to provision functional server space to users in 24 hours. The cost for a user to obtain an environment on RACE is reasonable and can be set up with an approved Government credit card.

According to DISA, personnel can expect the same level of service and availability when using RACE over a traditional environment. Additionally, for security purposes RACE has built-in application separation controls so that all applications, databases and web servers are separate from each other. DISA also has a strict data cleansing process for when an application needs to be removed completely from the RACE platform. Since the inception of this cloud-based solution, hundreds of military applications including command and control systems, convoy control systems, and satellite programs have been developed and tested on RACE.

Project: Forge.mil (Defense Information Systems Agency)

Typical implementation of new software and systems at DoD requires large amounts of time and money due to licensing, acquisition, and support demands. Non-cloud-based software development does not typically allow for the utilization of economies of scale, ubiquitous delivery, or cross-collaboration on projects. Recognizing that such benefits can be found in the cloud, DISA established the software development environment: Forge.mil. Through Forge.mil, DISA provides the entire Department of Defense with the tools and services necessary for rapid development, testing, and deployment of new software and systems.

Forge.mil teamed with cloud provider CollabNet to provide for a software development platform to allow users to reuse and collaborate on software code. Currently, Forge.mil has over 5,000 users, with over 300 open source projects, over 500 file release posts, and over 30,000 downloads. Forge.mil's collaborative environment and open development platform allow DISA to avoid large start-up costs and enable additional return on investment (ROI) through software reuse.

With rapid project start-ups at minimal cost, Forge.mil estimates new projects developed in its environment save DISA between $200,000 and $500,000 per project. Also, DISA estimates about $15 million in cost avoidance by utilizing an open source philosophy that allows for software reuse and collaborative development.

Tips

This open source philosophy of Forge.mil not only saves money on licensing and support, but provides improved software by giving version control, traceability, and having multiple stakeholders from various projects work on the same software code.

Forge.mil hosts an array of projects for different areas of DoD including the Army, Navy, Air Force, Marine Corps and the Joint Chiefs, all within a secure environment that appropriately protects DoD software assets. Forge.mil allows DISA and its customers to reduce their costs and shorten the time required to develop new software and systems by using a cloud environment that promotes collaboration, reuse of developed software, rapid delivery, and shortened time-to-market for projects.

Project: Personnel Services Delivery Transformation (PSDT) (United States Air Force)

Faced with a mandate to reshape the personnel community, the Air Force Personnel Center needed to reduce the time spent searching for documentation and allow personnel to support war-fighting missions. The Air Force Personnel Center created a program to transform the way human resource tools and services were delivered. The primary goal was to create a better customer experience by providing self-service solutions and tracking customer service needs.

The Air Force implemented the Software as a Service (SaaS) solution by RightNow to support its knowledge management, case tracking, contact center tracking and customer survey mission needs. Using tools available in the RightNow solution, the Air Force focused on solving fundamental problems with the way information was organized.

RightNow empowered the Air Force to complete its manpower reduction initiative and save over $4 million annually. Searches on the knowledge base have increased to nearly 2 million per week, and customer engagement has increased by 70 percent. By using a cloud-based solution, the site has been able to scale to meet fluctuating demand without compromising the customer experience. Customers can now find answers from over 15,000 documents within two minutes, an improvement on the twenty-minute wait they faced before the implementation of this solution.

Department of Energy

Project: Cloud Computing Migration (Lawrence Berkeley National Labs)

The Department of Energy is exploring cost and energy efficiencies that can result from leveraging cloud computing. This initiative explores how to use cloud computing to address needs across the enterprise, in specific business services, and in scientific study. Although started in 2009, these efforts at Lawrence Berkeley National Labs (LBL) are already showing promise.

LBL has already deployed over 2,300 mailboxes on Google Federal Premier Apps, and will end up with 5,000 e-mail accounts deployed by August 2010. This solution uses a LBL Identity

Management System to provide authentication. Additionally, Google Docs and Google Sites have already been deployed and are being used by small and medium-sized scientific research teams to foster collaboration and community documentation.

Presently, LBL is evaluating the use of Amazon's EC2 to handle excess capacity for mid-range computers during peak usage periods. LBL is also investigating the use of a federated identity to provide access for the scientific community to a wide range of cloud computing offerings. LBL estimates they will save $1.5 million over the next five years in hardware, software and labor costs from the deployments they have already made.

Department of Health and Human Services

Project: Supporting Electronic Health Records

The Department of Health and Human Services (HHS) is leveraging cloud computing to support the implementation of Electronic Health Records (EHR) systems. HHS is planning for 70 Regional Extension Centers which will assist over 100,000 Primary Care Practitioners. To coordinate healthcare providers' implementation of new EHR systems, HHS is deploying a cloud-based customer relationship and project management solution provided by Salesforce.com. The solution will support HHS's Regional Extension Centers in the selection, implementation, and meaningful use of EHRs. Various implementation approaches can be analyzed to quickly identify best practices for EHR implementation as they emerge.

After reviewing internal and cloud-based solutions, the Office of the National Coordinator (ONC) decided that Salesforce.com offered the best CRM solution for a quick, inexpensive, and rapidly scalable implementation. The review process concluded that it would have taken over a year to implement an internally-based system. Leveraging the cloud solution, ONC was able to implement the first phase of the Salesforce.com solution in less than three months after the award.

Remember

One of the advantages ONC anticipates from deploying a cloud-based CRM system is the ability to update the system as Regional Extension Centers start using it. More implementation phases are already planned to ensure that users' needs are met. ONC expects to be able to quickly update future phases of the system in substantially less time, while doing it collaboratively with end users.

Department of the Interior

Announced Project: Agency-wide E-mail

The Department of the Interior is pursuing a Software as a Service (SaaS) cloud computing model for e-mail. DOI has 80,000 e-mail users who are widely dispersed across the United States. They are currently supported by a very complex messaging infrastructure comprised of more than a dozen different e-mail systems. The department had already determined that a single e-mail

infrastructure would reduce the complexity of the overall system and improve the level of service provided to their users when it decided to explore cloud-based solutions.

Remember

When considering how best to deliver a single e-mail system, the department analyzed the opportunities for cost savings presented by cloud computing. The numbers were compelling: by implementing e-mail using an external commercial SaaS model, the department expects to provide improved service to its 80,000 users for one-third the amount of money that it spends today. The department is moving forward with this project with a completion date in FY 2011.

General Services Administration

Project: USA.gov (Office of Citizen Services)

As the Federal Government's primary information portal, USA.gov, presents the American people with a vast body of information and resources including topics like benefits and grants, taxes, jobs, education, health, voting, technology, and business and non-profit guides.

As the Federal Government encourages citizens to become more involved and active with local, state, and federal politics, key sites like USA.gov see vastly increasing and decreasing website traffic as key issues are debated in the national public forum, natural disasters come and go, and voting season approaches. These spikes in traffic made a cloud computing-based solution very attractive, as a cloud infrastructure is much better able to deal with on-demand scalability than most traditional IT infrastructures. This increased flexibility positions USA.gov to better serve emerging needs.

Remember

By moving to Terremark's Enterprise Cloud service, the General Services Administration (GSA) reduced site upgrade time from nine months (including procurement) to a maximum of one day. Monthly downtime moved from roughly two hours with the traditional hosting setup to near zero with the cloud solution (99.9 percent availability). With its legacy setup, GSA paid $2.35 million annually for USA.gov, including total hardware refresh and software relicensing costs of $2 million, in addition to personnel costs of $350,000. By moving to a cloud service, GSA now pays an annual total of $650,000 for USA.gov and all associated costs, a costs savings of $1.7 million, or 72 percent.

Announced Project: Agency-wide E-mail

GSA's current environment lacks the level of integrated features commercially available. GSA requires a greater use of features such as integrated messaging and collaborative tools to support its mission. E-mail archiving is currently implemented inconsistently, is difficult to use, and does not meet information retrieval (e-discovery) requirements. The storage associated with e-mail archiving continues to grow and is costly to manage. Recent regulations for handling e-mail litigation hold and discovery demand that GSA implement a more effective and expedient process. Additionally, GSA is seeking a solution that will reduce its in-house system maintenance burden and provide GSA users with more timely implementations of new versions and features.

GSA's e-mail effort will migrate over 15,000 mailboxes to a cloud-based solution, eliminating the redundant and disparate infrastructure presently housed at 17 different locations around the world.

Although still in the information gathering phase, initial estimates indicate that over the first two years, GSA will realize a 30 percent cost savings.

National Aeronautics and Space Administration

Project: World-Wide Telescope (Ames Research Center)

Nebula, NASA's cloud-computing platform, is helping NASA to engage the public through the viewing and exploration of the Moon and Mars in unprecedented resolution. Nebula allows NASA to process, store and upload thousands of high-resolution images and over 100 terabytes of data. In a traditional IT environment, it would have taken several months to procure new infrastructure and another one to two months of full-time work by two full-time employees to configure the new equipment to handle this data. By utilizing Nebula, NASA saved four to five months of time and roughly 800 hours of labor, allowing the agency to focus on expanding the content accessible to the public instead of building IT infrastructure.

The nature of NASA's activities requires strict security policies, creating a challenge in providing a collaborative environment to share data with outside partners or the public. Nebula's architecture is designed from the ground-up for interoperability with commercial cloud service providers, offering NASA researchers the ability to port data sets and code to run on commercial clouds. Nebula provides a secure way for NASA to make its data accessible to partners, avoiding the need to grant access to internal networks. Each researcher needs a varying amount of storage space and compute power to process his or her data sets. In the old operational model, these resources took months to procure and configure and required constant monitoring and frequent upgrades. Using Nebula's cloud computing infrastructure, researchers will be able to provision these services in just a matter of minutes.

Remember

NASA space exploration missions can take over 10 years to develop and the resources needed to process the data coming back are usually scheduled and procured well before launch. Missions, however, have a varying degree of success: some are delayed at a late stage, some are cancelled altogether, and some last much longer than originally anticipated. Nebula's cloud services allow NASA to be much more flexible and responsive to actual mission needs, scaling resources up or down as the actual requirements of the mission develop. In addition to supporting NASA's missions, the Nebula cloud computing platform has demonstrated additional versatility and has become the home of the Federal Government's flagship transparency website USAspending.gov. USAspending.gov 2.0 was completely reengineered to leverage the cloud-computing platform at Nebula, and growing the amount of storage as Federal spending data grows will now be a quick and easy process.

Project: Be A Martian (Jet Propulsion Laboratory)

NASA's Jet Propulsion Laboratory (JPL) brings science to the American people by inspiring interest in the planet Mars. The laboratory sought to increase the impact of its education and outreach program by using technology. It wanted not just to give Mars data to the public, but also to excite the public about Mars.

To meet this challenge, JPL developed an interactive website, BeAMartian.jpl.nasa.gov, using the Microsoft Azure cloud computing platform. An application programming interface (API) connects website visitors with 250,000 pictures of Mars, available without having to store any additional data on JPL computers. On the cloud, individuals can virtually explore the planet by browsing pictures, watching videos, and creating tags. They can post questions, read responses, and send messages to Mars. The more content a visitor contributes to the site, the more reputation points they earn in their account. For participants, this is a fun way to learn more about Mars.

"JPL chooses to keep it real through early exploration of multiple clouds," said Tom Soderstrom, Chief Technology Officer of NASA's JPL. "In other words, JPL wants to be an intelligent user of clouds and the only way we can do that is by being proactive and trying them out, end-to-end with real mission data. We've been exploring the clouds by partnering with JPL missions and industry partners for about two years now and have several very good use cases and stories."

Remember

With this cloud computing solution, NASA has successfully engaged a crowd of users. Users have created over 2,000 pieces of social media and inspired 200 stories on TV, radio, and in print. There have been 2.5 million API queries from NASA crowd-sourcing applications and 500,000 API queries from developers. The Town Hall area of the website has received over 40,000 votes and 5,000 individuals and teams have registered for a NASA sponsored competition. This crowd has also helped NASA identify craters and other features of the Martian surface. JPL has benefited from this outreach by having engaged users and by exploring and learning about new cloud-based technologies.

Announced Project: Enterprise Data Center Strategy

NASA recently announced that it is re-evaluating its enterprise data center strategy and has halted a request for proposals that would have yielded an indefinite delivery/indefinite quantity contract with a maximum value up to $1.5 billion for outsourced data center services over multiple years.

Concurrently, a number of organizations within NASA are evaluating the use of Nebula, NASA's scientific cloud solution for possible application in satisfying their mission data center needs:

The Flight Vehicle Research and Technology Division at Ames Research Center is exploring using Nebula for their Message Passing Interface (MPI) implementation. This group performs

flight vehicle air flow computation. Data from each piece of the aircraft surface runs on a different compute node and each node communicates edge conditions to its neighboring nodes using MPI. Currently, it takes a very expensive suite of equipment to do that work: NASA's 60000-core Pleiades computer. Although Nebula does not compete on performance with Pleiades, the setup time and money saved by self-provisioning computing power makes Nebula an attractive alternative.

A second mission organization with enormous memory and storage requirements is interested in Nebula because the Infrastructure as a Service (IaaS) beta version will allow them to specify the amount of memory and storage needed for their virtual machines. One of the group's storage-heavy applications requires 12 GB of memory, which can be accommodated on the Nebula IaaS cloud solution.

A third organization is evaluating Nebula to create virtual workstations for software developers to write and test-compile their code. Nebula would give them more fine-grained control over the development environment and allow developers to share the many modules and libraries currently running on their local desktops.

And yet another organization is evaluating Nebula as a service platform for interaction with non-NASA partners. Nebula would enable anonymous but controlled FTP for large file transfers and run an in-house, web-based java application that analyzes and visualizes data produced by NASA's Airspace Concept Evaluation System.

Social Security Administration

Project: Online Answers Knowledgebase (SOASK)

The Social Security Administration (SSA) handles millions of questions and inquiries from citizens every year. For example, inquirers want to know what they can do online, or how to get a social security number, file for benefits, locate a field office, get a retirement estimate, or request a proof of income letter. In order to provide the public with a convenient means to answer to their questions, anytime and anywhere internet access is available, the agency provides an online database of Frequently Asked Questions (FAQ).

The SSA is leveraging a cloud-based solution from RightNow Technologies to provide this service. Visitors to Socialsecurity.gov can search for answers by category, keyword or phrase, which helps them quickly find the information they are looking for. Over a thousand questions and answers are included in the knowledge base. SSA keeps the information contained in the knowledgebase up-to-date and relevant, eliminating the need to call or visit SSA for basic information.

In 2009, the number of answers provided through SSA's Frequently Asked Questions grew to over 34 million. Given current agency staffing levels, it would not have been possible for

office staff and 800-number agents to answer even 10 percent of these additional inquiries. By contrast, the internet solution is highly scalable, allowing SSA to meet increasing demand for online information without impacting service in the office and on the phone.

Federal Labor Relations Authority

Project: Case Management System

The Federal Labor Relations Authority (FLRA) recognized that its decade-old case management system was not supporting its mission to the fullest extent possible. FLRA's users regularly experienced delays in searching and the system couldn't keep up with expected growth. Additionally, the internal system had expensive software licensing costs.

Strategically, FLRA wanted to implement a shared electronic case management tracking system that would allow citizens to file cases and obtain documents electronically and then check the status of their cases. By using the cloud, FLRA intended to improve infrastructure and make existing IT and operations support more responsive to business needs while meeting regulatory compliance.

The FLRA selected Intuit's Quickbase system as its platform to implement this new system. From requirements-gathering to completed development, the project took less than 10 months to implement. The cloud-based solution has provided FLRA with more rapid development at 25 percent of the original time to deploy. Users now use a modern browser-based user interface, and information collaboration capabilities have improved work efficiency. FLRA estimates that the total cost of ownership of its case management system has been reduced by nearly $600,000 over five years.

Recovery Accountability and Transparency Board

Project: Recovery.gov Cloud Computing Migration

Launched in February of 2009, after the passage of the American Recovery and Reinvestment Act (Recovery Act), Recovery.gov is designed to "foster greater accountability in the use of funds made available by this Act." The Recovery Accountability and Transparency Board created this public-facing site to allow citizens to track how stimulus funds are spent. The site includes a number of tools including graphs, charts, and maps which are continuously updated and refined to properly reflect stimulus spending. As such, a Government-wide system relies on an agile and substantial infrastructure to ensure that information is accessible, secure, and easy to update with current information.

On April 26, 2010, Recovery.gov became the first Government-wide system to migrate to a cloud-based environment. The Amazon EC2 infrastructure will provide added security, as the vendor's security will supplement existing measures previously put in place by the Board. The elastic nature of this commercial cloud system means that Recovery.gov is a fully scalable site,

ready to handle spikes in usage as needed. In-house personnel currently dedicated to management of the site's associated data center and corresponding hardware will be able to redirect their resources to oversight and fraud detection.

Moving Recovery.gov to the cloud means a projected cost savings of $334,800 in FY 2010 and $420,000 in FY 2011. This represents 4 percent of the Board's $18 million total budget provided by Congress. Additionally, the Board plans to reallocate more than $1 million worth of hardware and software to its accountability mission to help identify fraud, waste, and abuse. Relocating Recovery.gov to the cloud ensures nearly 100 percent uptime and the ability to continuously backup site information. By implementing cloud technologies, the Board better meets its obligations laid out under Section 1526 of the Recovery Act, and is able to refocus efforts on its mission of transparency and accountability.

Securities and Exchange Commission

Project: Investor Advocacy System

The Office of Investor Education and Advocacy (OIEA) serves individual investors who complain to the SEC about investment fraud or the mishandling of their investments by securities professionals. The staff responds to a broad range of investor contacts through phones, e-mail, web-forms, and U.S. mail with volumes close to 90,000 contacts annually. Case files were previously tracked in a 10 year old in-house system. Like many older systems, there were several limitations including the inability to attach documents, handle paper files, and provide accurate reports. The older system was also intermittent in regards to up-time and system speed.

To address these issues, the SEC implemented a cloud-based CRM tool called Salesforce.com. The implementation of Software as a Service (SaaS) solution that took less than 14 months from inception to deployment. Since the implementation of OIEA, the SEC has realized improvements in system reliability, efficiency, and accuracy. Paper files are scanned into the system and worked electronically. All investor contact channels (e-mail, web-form, U.S. mail, fax, and phone) are brought into a single queue to be assigned and worked electronically. All documentation can now be attached to case files, which allows staff members to build complete chronology of events.

Using this new paperless environment, the time required to complete files has significantly been reduced. In many cases it was decreased up to 75 percent. Lifecycle tracking is now also available, allowing management the ability see at what stage and the chain of events for every case file. The system now also tracks information that is useful for assisting investors as well as reporting on data that is valuable to other SEC divisions.

Having this new solution better equips SEC in assisting investors efficiently and accurately, which is even more important as we are still dealing with the financial crisis.

State and Local Cloud Computing Case Studies

Introduction

The following case studies provide recent examples of how state and local governments are using cloud computing technologies.

Use cases

- ☐ State of New Jersey (New Jersey Transit Authority) - Customer Relationship Management
- ☐ State of New Mexico (Attorney General's Office) - E-mail and Office Productivity
- ☐ Commonwealth of Virginia (Virginia Information Technologies Agency) - Application Development Platform
- ☐ State of Wisconsin (Department of Natural Resources) - Collaboration
- ☐ State of Utah (Department of Technology Services) - Cloud Computing Services
- ☐ City of Canton, Georgia- E-mail
- ☐ City of Carlsbad, California - Communication and Collaboration Services
- ☐ City of Los Angeles, California - E-mail and Office Productivity
- ☐ City of Miami, Florida - 311 Service
- ☐ City of Orlando, Florida - E-mail
- ☐ Klamath County, Oregon - Office Productivity
- ☐ Prince George's County, Maryland - School District E-mail
- ☐ State of Colorado (Office of Information Technology) - Launching an Enterprise Cloud
- ☐ State of Michigan (Department of Technology Management and Budget) - MiCloud

State of New Jersey

Project: Customer Relationship Management (New Jersey Transit Authority)

NJ Transit is the nation's largest statewide public transportation system providing bus, rail, and light rail services of over 900,000 daily trips on 247 bus routes, 26 bus stations, 11 commuter rail lines, and 3 light-rail lines. NJ Transit links major points in New Jersey, New York, and Philadelphia, serving 164 rail stations, 60 light-rail stations, and 19,800 bus stops. NJ Transit relies upon its ability to field and respond to customer feedback, and requires a robust customer service system. The agency transformed its customer feedback process from one where issues went unresolved, with no tracking and, in some cases, with multiple executives seeing and responding to the same inquiry, to a streamlined, faster, more accurate, and more efficient response system. The legacy infrastructure for tracking customer information and inquiries had limited functionality and not all customer inquiries were properly documented for future use. In addition, customer service

representatives were responsible for a wide variety of inquiries, limiting the depth of knowledge they could apply to any given inquiry.

When NJ Transit began the search for a new customer system, the organization found that a hosted CRM system from Salesforce.com service fit its needs. To take full advantage of the software's capabilities, NJT realigned its customer service department to make each member of the staff the expert for a specific customer service area, which decreased communications overhead and improved productivity. The cloud-based system provides workflow rules that route incoming customer questions to the subject area experts. It also enabled customers and internal users the ability to ask questions and submit issues on the existing site via an online "Contact Us" web form, which flows into the solution's central customer information warehouse. The system's applications are linked to a data warehouse, employee information, an e-mail management system, and a data quality system.

Under the new system, the same number of staff handled 42,323 inquiries in 2006, compared with 8,354 in 2004. During its use, and without an increase in staff, the average response time to inquiries dropped by more than 35 percent and productivity increased by 31 percent. The web form cut down on the time spent handling free-form e-mail; approximately 50 percent of all customer feedback is captured via the online form. Salesforce.com has allowed NJ Transit to make significant improvements in their customer service capabilities while simultaneously reducing cost, infrastructure, and response time.

State of New Mexico

Project: E-mail and Office Productivity Tools (Attorney General's Office)

The New Mexico Attorney General's Office (NMAGO) has nearly 200 full-time employees, including 120 attorneys. Most work in the office revolves around creating, storing, and accessing documents in a secure IT environment. The office had historically relied upon the case management functionality of its e-mail system to track official documents ranging from legal briefs to news releases. However, this system did not offer a secure back-up function, leaving vital and sensitive documents exposed to possible loss in case of a server failure. One possible solution, migrating to a system of in-house servers, was cost-prohibitive in the short- and long-term, with the upfront investment calculated at $300,000. For this reason, the office explored alternative IT upgrades.

When investigating alternative e-mail systems, NMAGO selected Google Apps Premier Edition to meet its needs. This solution provides the necessary backup capabilities, and the mail search function also eases the difficulty of locating specific files. Without the need for in-house hardware, employees now have an unlimited ability to access, save, and archive their e-mails and documents. This transition has also been beneficial from an environmental perspective, as it has

reduced the need for paper versions of files. NMAGO is now able to avoid costs such as the $50,000 spent a few years ago for replication software to store data to a disaster recovery site. The office has reduced costs and energy use through reduced hardware acquisitions and reduced workloads for IT staff. Additionally, the office has reduced time and money spent on licensing.

NMAGO's successful migration to the cloud is an example of what the office's CIO calls a "fundamental shift in the way information is transported to users." The CIO and his team believe that the sharing platform offered by cloud-based solutions is easily replicable and can be used to meet various government needs. They "believe cloud computing offers a new way for government to be more responsive and helpful to the public, and save more money than ever before."

Commonwealth of Virginia

Project: Enterprise Application Development Platforms (Virginia Information Technologies Agency)

The Virginia Information Technologies Agency (VITA) is the Commonwealth's consolidated information technology organization with responsibility for governance of IT investments and the security, technology procurement, and operations of IT infrastructure. The Agency Outreach group of the Enterprise Applications Division (EAD) at VITA provides software development and integration support and services to small agencies, the secretariat, and projects that require cross-agency collaboration.

In the spring of 2009 this group received a request from the Secretary of the Commonwealth to build and host an online solution for Notary and eNotary applications. At the same time it also received a request from Virginia's Workforce One Stop councils to create and host a low cost solution for a common intake form for the centers. Given the limited resources available, under the constraints of traditional technologies, these custom development projects would have been cost prohibitive.

To meet this challenge, EAD leveraged cloud computing to quickly provision a virtualized software development platform. Using Amazon EC2 services, the group was able to add and remove development or testing environments with ease. Even after full release to a production environment, EAD uses cloud computing to scale the production environment up as needed and for disaster recovery backups through RackSpace virtual storage services.

Remember

Virginia used cloud computing to develop new applications that would have otherwise been cost prohibitive. Instead of going through a process that typically lasts months, EAD was able to stand up a virtualized development platform from the cloud in less than two hours. VITA is still evaluating cloud computing at the agency level, reflecting on this group's recent success delivering service with the speed and flexibility its customers need.

State of Wisconsin

Project: Collaboration (Department of Natural Resources)

The Wisconsin Department of Natural Resources (DNR) has 200 locations scattered across the state, including some in remote areas. In the past, the department typically conducted business through conference calls and face-to-face meetings with staff from various locations. Outside of e-mail, there were no ad-hoc collaboration tools available to department staff to review documents which required multiple revisions by different staff members. The department's available video-conferencing system ran using outdated technology and cost the DNR $1,330 per month.

The DNR evaluated server-based collaboration software, but due to a recent migration of all of the department's servers to the state's new data center, there were no resources available to purchase an on-premises solution. The DNR began using Microsoft Live Meeting as a web conferencing solution and immediately realized cost savings and improved efficiencies. Staff members are now able to interact and can use dynamic solutions including a 360-degree panoramic video camera to conduct meetings together. The cost of running a web conference is only a fraction of the cost required to use a traditional conference call bridge and the DNR has the flexibility to purchase additional user licenses as needed for other departments.

The DNR has used this cloud-based solution to completely replace on-site meetings, training, and telephone and conference calls among internal staff and with outside agencies. This solution allows remote users to participate in meetings even if they are not at one of the DNR's office locations. The staff is now more efficient because they spend significantly less time on the road travelling to meetings. Since this migration in 2009, the DNR estimates that staff members have participated in nearly 3,500 meetings, saving the department more than $320,000. In the coming years, the DNR expects the return on investment to grow from 270 percent for the first year to over 400 percent in future years.

State of Utah

Project: Cloud Computing Services (Department of Technology Services)

The State of Utah began an effort to standardize and unify its IT services in 2005 when it merged its technology assets into a single department, moving all IT staff under the state Chief Information Officer. To develop a suitable cloud strategy, the state needed to fulfill specific security requirements unique to the state. If these security challenges were met, Utah could take advantage of an array of cloud benefits including elastic expansion of services, rapidly provisioned computing capabilities, and shared services across multiple users and platforms based on customer demand.

After a wave of data center consolidation, in 2009 Utah decided that a hybrid cloud approach would work best for the state's needs. A hybrid approach combines access to public cloud services that add to or replace existing state infrastructure with private cloud services meeting specialized

access and security requirements. This cloud environment includes services hosted both by third-party providers and in-state data centers.

The move to cloud-based solutions has benefited local public sector actors across the state. Although many of the public cloud provisioned services are free, the State of Utah also supports a growing number of paid services where individual county and city governments pay only for their usage. These currently include Force.com for Customer Relationship Management, Google Earth Professional for shared Geographic Information System (GIS) planning, and Wikispaces where there are an increasing number of self-provisioned wikis. Contracts for these services are centrally managed through the Department of Technology Services (DTS), and make it easy for agencies to use.

Having provided its agencies and local governments with centralized access to the public cloud, the State of Utah is now focused on completing its private cloud. The state's applications previously resided on about 1,800 physical servers in over 35 locations. By December 2010, the State planned to move these applications to a virtual platform of 400 servers. This initiative is expected to save $4 million in annual costs for a state IT budget of only $150 million. Going forward, DTS plans to extend virtualization to desktops across the state.

Remember

By implementing a vast strategy for migrating services to the cloud, the state has created an enterprise where public or private services can be reused and provisioned on demand to meet agency needs as cost-effectively as possible. This effort has had an immediate impact on state agencies and is expected to result in significant future savings.

City of Canton, Georgia

Project: E-mail

The City of Canton, GA, approximately 40 miles north of Atlanta, has a population of 21,000, and was recently recognized as the fastest growing city in Georgia and America's 5th fastest growing city. The city's 185 employees were looking to reduce the cost and effort of maintaining an IT infrastructure and increase the reliability of business e-mail and productivity solutions. The legacy e-mail system was difficult to maintain and keep up-to-date.

The city decided to migrate to Google Apps to provide staff members with a more reliable and feature-rich system. Staff members immediately benefited from gaining access to e-mail at home and on mobile devices, and from the increased collaboration available with tools such as group calendar, instant messaging, and shared documents and spreadsheets.

With this cloud-based solution, the city's IT staff no longer has to handle spam filters, a task that took 20 hours a week to manage prior to migration. IT staff members are also able to use powerful e-mail discovery tools in the new cloud-based solution so that e-mails potentially related to legal investigations are securely archived but easily accessible to approved personnel. The city estimates an annual savings of $10,000 by migrating to cloud computing.

Tips

City of Carlsbad, California

Project: Communication and Collaboration Services

The City of Carlsbad, California has 1,100 employees across 22 departments who work in 30 different facilities across the city. Until recently, the city's employees used a 15 year-old, non-standard IT system. The city's IT department knew it had to simultaneously standardize its IT infrastructure and cut costs without sacrificing its high level of service. When the city began the process of standardizing its IT infrastructure, officials decided to review options for migrating from an on-premise e-mail and collaboration system to one hosted in the cloud.

The city ended up selecting a cloud-based version of the Microsoft productivity suite, hosted in Microsoft's data centers outside of Carlsbad, CA. It was able to eliminate the costs of maintaining equipment, paying only monthly user fees for this new environment. While the city considered using an on-site version of the productivity suite instead due to data security concerns, it realized that from a disaster recovery standpoint, their data was more secure being hosted outside of the city's data center.

The city has already realized a 25 percent savings over the past year using the new off-site solution, as there is no longer a need to maintain servers, manage upgrades, implement hardware replacements, or hire a systems administrator. The city realized other immediate benefits after the migration, including better access to e-mail from mobile devices and new, integrated instant messaging and web collaboration for meetings and video conferences.

Remember

City of Los Angeles, California

Project: E-mail and Office Productivity

The City of Los Angeles has 34,000 employees across 40+ departments. In 2009, the city faced a $400 million deficit. This budgetary crisis and the resulting IT staff layoffs exacerbated existing frustrations with the city's in-house IT systems. The city's Information Technology Agency sought to find a new e-mail and IT provider which would streamline productivity and create more efficiencies in day-to-day operations. The city received 15 proposals for possible replacements to its in-house system.

In October 2009, the City of Los Angeles announced plans to transition to Google Apps with the help of systems integrator, with a five year services contract. The city planned on having all employees on cloud-based e-mail by June 30, 2010 and has begun initial use of other products within the Google Apps Premier Edition suite and to the cloud as city employees have become more familiar with using the cloud for workplace productivity.

The city's Chief Technology Officer estimated a direct savings of $5.5 million over five years as a result of the implementation, with the potential ROI for increased productivity possibly reaching $20 to $30 million as city employees become fully trained on cloud-based applications. The city is now able to offer each city employee 25 times more storage space, can provide much more capability, and add new users without ever needing to worry about hardware availability on city servers. City employees will also benefit from cloud-based integrated instant messaging, video conferencing, simultaneous review and editing of documents by multiple people, and the ability to access their e-mail and work data from any computer or mobile device.

Remember

While some city council members and staff were skeptical at first about moving city data outside of direct city control, the vendors have ensured that from a security and disaster recovery standpoint, data being stored in the cloud environment exceeds both the city's contractual requirements and current environment. The city's new system and its data will be safe from earthquakes and other potential natural disasters that could and have affected the city. In addition, the City of Los Angeles retains full ownership of all data on the servers and the vendors must request access to see city data, stored in the cloud. These were critical hurdles the system had to clear before being recommended by staff and accepted unanimously by the city council. With these protections and the productivity benefits, moving to cloud computing was a natural step for Los Angeles and in keeping with their focus on innovation as well as fiscal responsibility.

City of Miami, Florida

Project: 311 Service

The City of Miami, with a population of nearly 5.5 million, has 3,600 employees who work in 83 locations. When the city's centralized IT department needed to cut its budget by nearly 18 percent, and was forced to drop nearly 20 percent of its already small staff, continuing to deliver quality and innovative services became a challenge. At the same time, the city sought to supplement its 311 phone line, used by citizens to report non-emergency situations, with an interactive online platform for tracking service requests and mapping them geographically.

The 311 website proposal posed several challenges to the city and its IT staff. The city needed to be sure it had adequate processing power to support its new, processing power-intensive mapping application. The city also needed to take into account disaster recovery measures, since the Miami area is frequently hit with hurricanes. Overall, the city was unsure it could provide the necessary resources to manage the 311 website in-house, so moving to the cloud was the logical next choice.

Remember

The city decided to leverage a scalable, cloud-based Windows Azure platform that provides developers with on-demand hosting in Microsoft data centers. From a technical standpoint, the city was able to seamlessly integrate existing technologies in use by development teams on other projects with the cloud-based platform. Also, the pay-as-you-go platform allowed the city to test out the application and only pay for actual usage, which was also beneficial when the application becomes more popular. Moreover, IT staff members were able to streamline development of the application and move from testing to production simply and quickly. The deployment of the 311 website application on the cloud-based platform was successful and the city is planning additional service offerings to citizens based on the overall value and efficiency of cloud computing.

City of Orlando, Florida

Project: E-mail

To address recent budget and human resource challenges, the City of Orlando moved its e-mail and productivity solution to the cloud. Preparing for the FY 2010, the city faced a 12 percent budget cut and the retirement of two mail administrators and an information security officer. As the license renewal deadline approached, the city's CIO confronted these business challenges by leading Orlando into the cloud.

After evaluating several providers, Orlando chose to contract with Google to provide an e-mail solution for all 3,000 city workers. City leadership supported the transition based on several decision factors including projected cost savings of $262,500 per year, centralized document storage and collaboration, increased mail storage from 100MB to 25GB per user, and enhanced support for mobile devices.

Although the city's contract includes Google Docs, the city also retained the Microsoft Office productivity suite in order to avoid the cost to retrain employees.

After piloting with a small number of users, the full roll-out of the solution occurred on January 7, 2010. The city has realized a 65 percent reduction in e-mail costs and provided additional features to increase the productivity of workers. Google is now responsible for the city's e-mail server maintenance and IT support. Security functions and features such as virus checking and spam control are also performed by Google through their Postini services.

Klamath County, Oregon

Project: Office Productivity

Klamath County, Oregon is one of Oregon's geographically largest counties spanning 6,000 square miles. The county has about 70,000 residents and a staff of 600 employees spread across the expansive county. County employees typically leverage video conferencing on a regular basis. When the county's director of IT realized that the need to upgrade IT capacity was imminent,

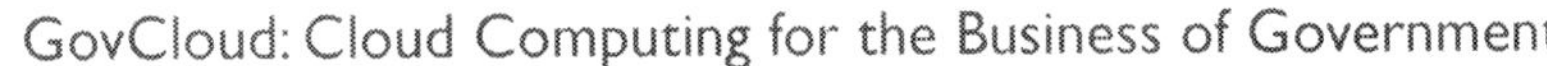

coupled with the fact that the county faced a budget crunch, he began evaluating cloud-based solutions.

After considering the options, the county decided to migrate to a hosted solution, and selected Microsoft Online Business Productivity Suite. This would not only free up valuable human resources from managing the server environment, but also cut costs. The county also noted the potential for dependability of the system to improve since performance was not tied to county IT staff's ability to keep the servers optimized.

Remember

With this migration, the county was able to keep costs low and ensure that IT personnel and other resources were used appropriately amidst the flat county revenues. The personnel required to manage the e-mail solution decreased by 1.5 full time equivalents, an 86 percent reduction. The county also managed to implement new features including integrated messaging, collaborative tools to increase productivity for the entire county, and the ability to archive e-mails for a longer period of time.

Prince George's County, Maryland

Project: School District E-mail

The Prince George's County, MD school district is the 18th largest school district in the country, with over 200 schools, 129,000 students, and nearly 28,000 faculty/staff. For the 2008-2009 school year, the school district was facing budget cuts of $185 million and projected that a needed upgrade to their on-premises e-mail system would cost $1 million. The existing system required the support of multiple dedicated members of the district's IT staff, and due to the lack of an e-mail archiving system, IT personnel spent an excessive amount of time tracking down electronic records for legal purposes.

Remember

The district decided to migrate staff e-mail accounts to the Google Apps platform, which is offered to public schools at no cost. The school district's faculty and staff are now leveraging Google's cloud computing platform for messaging and collaboration. More than 13,000 of the staff members also use Google Message Discovery, powered by Postini, for archiving and discovery. Due to the cost effectiveness of the cloud computing solution, the district was able to also add the Message Discovery add-on, which costs only a few dollars per user per year, allowing authorized users to locate e-mail messages within minutes. With the success of this cloud computing experience the school district is also considering phasing in a cloud-based solution for use by students throughout the school district.

State of Colorado

Announced Project: Launching an Enterprise Cloud (Office of Information Technology)

In 2008, Colorado's Governor's Office of Information Technology (OIT) began to consolidate the IT systems from 17 executive branch agencies. Prior to consolidation, the state was responsible for 40 data centers consisting of 1,800 servers, of which 122 alone powered different versions of

Lotus Notes, Microsoft Exchange, and Novell GroupWise for e-mail. The goal of consolidation was to achieve cost savings through standardization while reducing the complexity of administering multiple platforms, and improving service delivery. OIT also envisioned gaining the ability to share resources with local jurisdictions and schools across the state.

Colorado decided to implement a hybrid cloud solution to meet the diverse needs of its 17 state agencies. Each agency has its own applications which required different levels of security, so the state's plan includes three elements: a private cloud for line-of-business/highly secure data and systems, a virtual private cloud for archival storage/disaster recovery, and a public cloud for e-mail office productivity applications and websites.

For Colorado's private cloud, the state will use an existing data center and begin to leverage server virtualization. All production data will remain on-site, while virtualized instances of the production server can be stored off-site, increasing disaster recovery capabilities at reduced cost. Colorado's virtual private cloud allows for additional scalability on a pay-as-you-go model for large systems. Colorado has recently started transitioning systems to the virtual private cloud.

Colorado's usage of the public cloud will initially be a pilot of Google Apps for e-mail and office productivity. Using cloud-based e-mail provides Colorado with increased mobility, disaster recovery, storage, better document sharing, and collaboration. The pilot will test the migration of e-mail from three different agencies, focusing on security and workflow processing. If the pilot is successful and the cost-benefit analysis proves positive, the state plans to transition all 27,600 executive branch employees to the new system.

By shifting e-mail to the cloud, Colorado will be able to take all 122 existing e-mail servers out of production and experience significant operational cost savings. An initial cost-benefit analysis of the migration to cloud-based e-mail estimates annual savings of $8 million. In addition, Colorado will avoid additional expenses of up to $20 million over the next three years.

Remember

State of Michigan

Announced Project: MiCloud (Department of Technology Management and Budget)

In March 2010, Michigan's Department of Information Technology consolidated with the state's Department of Management and Budget. The new Department of Technology, Management and Budget (DTMB) is now building a full array of services to provide across governments and the private sector. Michigan is moving toward leveraging cloud-based solutions to provide clients with rapid, secure, and lower cost services though a program dubbed "MiCloud."

One key area of current action is the state's strategic investment in storage virtualization technologies, expected to go live in October 2010. Michigan is actively piloting MiCloud "Storage for Users" and "Storage for Servers" as internal government cloud functions delivered by

DTMB. The consumption expectation is more than 250 terabytes in the first year of operation at a projected storage cost that is 90 percent lower than today's lowest-cost storage tier. MiCloud provides self-service and automated delivery within 10 minutes of submitting an online request. The following table expresses projected savings based on migration rates. It is important to note that this low-cost option represents a service alternative that is only appropriate for data that does not require 24x7 availability or real-time, block-level replication.

Remember

The State of Michigan's 2010-2014 strategic plan also outlines critical future investments in virtual server hosting and process automation. The state is in the proof-of-concept phase for the MiCloud "Hosting for Development" and "Process Orchestrator" functions in the internal government cloud. The hosting for development function automates the delivery of virtual servers within 30 minutes of submitting an online request. Michigan will also explore a hybrid cloud to deliver a more complex Application Platform as a Service (APaaS). The process orchestrator function enables agency business users, regardless of IT skill level, to create and test simple process definitions. Business users will be able to publish processes and related forms to the service catalog and over time analyze related metrics. Ultimately, the shift to cloud computing will allow Michigan to improve services to citizens and business while freeing up scarce capital, staff resources, and IT assets for critical investments.

Migration Rate	Potential Annual Savings or Cost Avoidance
10%	$228,000
20%	$456,000
30%	$684,000
40%	$912,000
50%	$1,140,000
60%	$1,368,000

REFERENCES

Air Force Office of the Chief Information Officer. May 2010.

City of Canton. May 2010.

City of Carlsbad. May 2010.

City of Los Angeles. May 2010.

City of Los Angeles Information Technology Agency. "Los Angeles Google Enterprise E-mail and Collaboration System." presentation, 2010.

City of Miami. May 2010.

City of Orlando. "Orlando Goes Google." government document, 2010.

Defense Information Systems Agency. May 2010.

Defense Information Systems Agency. May 2010.

Department of Energy Office of the Chief Information Officer. May 2010.

Department of Health and Human Services. May 2010.

Department of the Interior Office of the Chief Technology Officer. April 2010.

Erlichman, Jeff. "Cloud Recruiting." On the Frontlines: Shaping Government Clouds, (Winter 2010). http://www.mygazines.com/issue/5865.

Federal Labor Relations Authority. April 2010.

Feeney, Tom C. "NJ Transit to test online suggestion box for riders." Nj.com, (May 14, 2008). http://www.nj.com/news/index.ssf/2008/05/nj_transit_to_test_online_sugg.html.

General Services Administration. May 2010.

General Services Administration. "Cloud Sourcing Models." government document, 2010.

General Services Administration. "FDCCI – Initial Data Center Inventory." government document, 2010.CIO Council. http://www.cio.gov/documents_details.cfm/uid/25A781B7-

BDBE-6B59-F86D3F2751E5CB43/structure/OMB%20Documents%20and%20Guidance/category/Policy%20Letters%20and%20Memos.

Google Enterprise Blog, The. http://googleenterprise.blogspot.com/2009/11/microsoft-exchange-or-google-apps-one.html.

Klamath County Oregon. May 2010.

LA GEECS Google Site, The. https://sites.google.com/a/lageecs.lacity.org/la-geecs-blog/home.

Lawrence Berkeley National Labs Office of the Chief Information Officer. May 2010.

National Aeronautics and Space Administration. March 2010.

National Aeronautics and Space Administration Jet Propulsion Laboratory Office of the Chief Technology Officer. May 2010.

National Aeronautics and Space Administration Office of the Chief Technology Officer. "WWT Case Study." government document, 2010.

National Aeronautics and Space Administration Office of Legislative and Intergovernmental Affairs. May 2010.

National Institute of Standards and Technology. http://csrc.nist.gov/groups/SNS/cloud-computing/.

National Institute of Standards and Technology. "Summary of NIST Cloud Computing Standards Development Efforts." government document, 2010.

New Jersey Transit. May 2010.

New Mexico Attorney General's Office of the Chief Information Officer. May 2010.

Prince George's County Public Schools. May 2010.

Prince George's County Public Schools. "Googlizing the Masses." presentation, School Board of Prince George's County Public Schools, MD, 2010. http://docs.google.com/present/view?id=dxjw4sx_14gvr3r7fz.

Recovery Accountability and Transparency Board. "Recovery.gov Moves to Cloud Computing Infrastructure." May 2010. http://www.recovery.gov/News/mediakit/Pages/PressRelease05132010.aspx.

Schlueb, Mark. "Orlando goes Google for cheaper e-mail." Orlando Sentinel, (January 2010). http://articles.orlandosentinel.com/2010-01-09/news/1001080262_1_google-e-mail-google-enterprise-google-docs.

Securities and Exchange Commission Office of Investor Education and Advocacy. April 2010.

Social Security Administration. May 2010.

State of Colorado Government Office of Information Technology. "Moving Colorado to the cloud: A business case." government document, 2010.

State of Michigan Department of Technology, Management & Budget. "Governing in the cloud – a government case study from Michigan." government document, 2010.

State of Utah. May 2010.

State of Utah Department of Technology Services. "Implementing Utah's cloud computing strategy: A case study on bringing cloud-based IT services to government." government document, 2010.

State of Wisconsin. May 2010.

United States Army G-1. May 2010.

U.S. Congress. American Recovery and Reinvestment Act of 2009. H.R. 1. 111th Cong., 1st sess. (January 2009). http://frwebgate.access.gpo.gov/cgi-bin/getdoc.cgi?dbname=111_cong_bills&docid=f:h1enr.pdf.

U.S. Congress. Energy Independence and Security Act of 2007. H.R. 6. 110th Cong., 1st sess. (January 2007). http://frwebgate.access.gpo.gov/cgi-bin/getdoc.cgi?dbname=110_cong_bills&docid=f:h6enr.txt.pdf.

Virginia Information Technologies Agency. "Cloud computing: Commonwealth of Virginia." government document, 2010.

Yasin, Rutrell. "City of Miami takes citizen services to cloud." Government Computer News, (March 2010). http://gcn.com/articles/2010/03/10/city-of-miami-microsoft-azure.aspx.

Cloud Resources

The following is a list of many of the leading cloud computing bodies and organizations.

Government Resources

Apps.gov

CIO Council

Information Security and Identity Management Committee (ISIMC)

Trusted Internet Connection (TIC)

NIST Cloud Computing Page

FedRAMP

Federal Data Center Consolidation Initiative

Cloud/Technology Resource Groups

Armed Forces Communications and Electronics Association (AFCEA)

Cloud Computing Interoperability Forum

Cloud Security Alliance (CSA)

Distributed Management Task Force (DMTF)

Internet Engineering Task Force (IETF)

International Test and Evaluation Association

Jericho Forum

Object Management Group (OMG)

Open Authentication (OATH)

Open Cloud Consortium (OCC)

Open Grid Forum (OGF)

Organization for the Advancement of Structured Information Standards (OASIS)

Storage Networking Industry Association (SNIA)

World Wide Web Consortium(W3C)

Other Resources

American Council for Technology (ACT) and Industry Advisory Council (IAC)

TechAmerica

Meritalk

http://kevinljackson.blogspot.com

http://govcloud.ulitzer.com

Glossary

Application-Centric: Description for any product or service that exists to support the needs/ requirements of an end user application – as opposed to "infrastructure-centric." For example, a product like Tivoli is infrastructure-centric while a product like UniCloud is application-centric, because it bases its automated decisions on application SLAs.

Cloud app: a software application that is never installed on a local machine — it's always accessed over the internet.

Cloud arcs: short for cloud architectures. Designs for software applications that can be accessed and used over the internet. (Cloud-chitecture is just too hard to pronounce.)

Cloud bridge: running an application in such a way that its components are integrated within multiple cloud environments (which could be any combination of internal/private and external/ public clouds).

Cloudcenter: a large company, such as Amazon, that rents its infrastructure.

Cloud client: computing device for cloud computing. Updated version of thin client.

Cloud Computing: While there are many definitions out there, the definition by which Univa operates is: A category of computing solutions where any "cloud" solution is a technology and/ or service that lets users access computing resources on demand (meaning: as needed) – whether resources are physical or virtual, dedicated or shared, and no matter how they are accessed (via a direct connection, LAN, WAN or the internet). A cloud environment must have four basic characteristics:

- ☐ Support boundless applications
- ☐ Ability to pool resources
- ☐ Service-based approach to application delivery
- ☐ Support for virtualization

Cloud Computing Services: cloud providers fall into three categories: software-as-a-service providers that offer web-based applications; infrastructure-as-a-service vendors that offer web-based access to storage and computing power; and platform-as-a-service vendors that give developers the tools to build and host web applications.

Cloud Enablement: The process of preparing one of the following, usually involving a service engagement and infrastructure/application analysis:

- ☐ A service provider to be able to deliver cloud-based services
- ☐ An end user company to be able to leverage cloud technology internally by building a private cloud environment
- ☐ An application to be able to run on a cloud infrastructure, whether internal or external (also called Cloud On-boarding or Cloudifying)

Cloud enabler: vendor that provides technology or service that enables a client or other vendor to take advantage of cloud computing.

Cloud envy: used to describe a vendor who jumps on the cloud computing bandwagon by rebranding existing services.

Cloud OS: also known as platform-as-a-service (PaaS). Think Google Chrome.

Cloud portability: the ability to move applications and associated data across multiple cloud computing environments.

Cloud provider: makes storage or software available to others over a private network or public network (like the internet.)

Cloud service architecture (CSA): an architecture in which applications and application components act as services on the internet

Cloud storage: You pay-as-you-go; the advantage is that while you will never own the technology, the provider will ensure that it is always up to date.

Cloudburst: an outage or security breach when your data is unavailable.

Cloud as a service (CaaS): a cloud computing service that has been opened up into a platform that others can build upon.

Cloud Operating System: A cloud operating system is a new category of software that is specifically designed to holistically manage large collections of infrastructure – CPUs, storage, networking – as a seamless, flexible and dynamic operating environment.

Cloud-oriented architecture (COA): IT architecture that lends itself well to incorporating cloud computing components.

Cloudsourcing: outsourcing storage or taking advantage of some other type of cloud service.

Cloudstorm: connecting multiple cloud computing environments. Also called cloud network.

Cloudware: software that enables building, deploying, running, or managing applications in a cloud computing environment.

Cloudwashing: slapping the word "cloud" on products and services you already have.

Content delivery network (CDN): A system consisting of multiple computers that contain copies of data, which are located in different places on the network so clients can access the copy closest to them

Cluster: A group of linked computers that work together as if they were a single computer, for high availability and/or load balancing.

Consumption-based pricing model: A pricing model whereby the service provider charges its customers based on the amount of the service the customer consumes, rather than a time-based fee. For example, a cloud storage provider might charge per gigabyte of information stored.

Disruptive technology: A term used in the business world to describe innovations that improve products or services in unexpected ways and change both the way things are done and the market. Cloud computing is often referred to as a disruptive technology because it has the potential to completely change the way IT services are procured, deployed, and maintained.

Elastic computing: The ability to dynamically provision and de-provision processing, memory, and storage resources to meet demands of peak usage without worrying about capacity planning and engineering for peak usage.

External Cloud (or Public Cloud): A cloud environment which exists outside a company's firewall, offered as a service by a 3rd party vendor (eg. Amazon EC2, Sun OCP, Google AppEngine).

Funnel cloud: discussion about cloud computing that goes round and round but never turns into action (never "touches the ground")

Grid computing: (or the use of a computational grid) is applying the resources of many computers in a network to a single problem at the same time - usually to a scientific or technical problem that requires a great number of computer processing cycles or access to large amounts of data.

HaaS: Hardware as a service; see IaaS.

Hosted application: An internet-based or web-based application software program that runs on a remote server and can be accessed via an internet-connected PC or thin client. See also SaaS

Hybrid Cloud (or Mixed Cloud): A cloud environment in which external services are leveraged to extend or supplement the internal cloud – simply put, a mixture of both private and public cloud.

IaaS: Infrastructure-as-a-service, or IaaS, is the category of cloud computing that refers to web-based access to storage and computing power on the cloud. It is also known as the 'Elastic Cloud' as the server capacity and application back-end scalability could be extended based on application's users demand. Included in this category are providers of infrastructure software needed to deploy private clouds based on virtualization technologies.

Internal Cloud (or Private Cloud): A cloud environment which creates a pool of resources behind a company's firewall and includes resource management and dynamic allocation, chargeback and support for virtualization.

Mashup: A web-based application that combines data and/or functionality from multiple sources.

Microsoft Azure: Microsoft cloud services that provide the platform as a service (see PaaS), allowing developers to create cloud applications and services.

Middleware: Software that sits between applications and operating systems, consisting of a set of services that enable interoperability in support of distributed architectures by passing data between applications. So, for example, the data in one database can be accessed through another database

Mixed Cloud (or Hybrid Cloud): A cloud environment in which external services are leveraged to extend or supplement the internal cloud – simply put, a mixture of both private and public cloud.

On-demand service: A model by which a customer can purchase cloud services as needed; for instance, if customers need to utilize additional servers for the duration of a project, they can do so and then drop back to the previous level after the project is completed.

PaaS: Platform as a service, or PaaS, is one of the categories of Cloud Computing; it delivers a fully baked application development environment you can subscribe to and use immediately; with PaaS, developers use free programming tools offered by the service provider to create applications and deploy them in the cloud. The infrastructure is offered by the PaaS provider or its partners, which charge by some usage metric such as CPU use or page views.

Pay-as-you-go: A cost model for cloud services that encompasses both subscription-based and consumption-based models, in contrast to traditional IT cost model that requires up-front capital expenditures for hardware and software.

Personal cloud: a personal wireless router that takes a mobile wireless data signal and translates it to wi-fi.

Pooling resources: The act of creating a single virtual resource from multiple physical or virtual resources – this pool has flexible internal boundaries and can be divided on demand to serve the needs of various customers and applications. For example, a service governor accesses a pool of resources to allocate any number or subset of these to meet the needs of a specific application. Service providers pool data center resources (physical and virtual) to create a flexible base of computing power that can be allocated to customers, in order to avoid the one-machine-one-customer (or x-machines-one-customer) paradigm, which wastes resources

Private Cloud (or Internal Cloud): A cloud environment which creates a pool of resources behind a company's firewall and includes resource management and dynamic allocation, chargeback and support for virtualization.

Public Cloud (or External Cloud): A cloud environment which exists outside a company's firewall, offered as a service by a 3rd party vendor (eg. Amazon EC2, Sun OCP, Google AppEngine).

Resource: Within the context of IT, resource refers to any item that can be used to support the needs of an application. Resources can be physical hardware, virtual machines, software systems, networks, storage, and even human

Roaming workloads: the backend product of cloud centers.

SaaS (Software as a Service): A model of software deployment where a provider licenses (or provides for free) an application to customers for use as a service on demand – examples include Salesforce.com or GMail. While falling out of fashion somewhat as a term, SaaS is still extremely relevant as the primary foundation on which cloud is based.

Service Migration: The act of moving from one cloud service or vendor to another.

Service Provider: The company or organization that provides a public or private cloud service.

SLA: Service level agreement – an agreement a vendor makes with an end user about the quality of the item being delivered. For application services, this would be an agreement to ensure a high quality of application performance, typically based on things like application availability and response times

Subscription-based pricing model: A pricing model that lets customers pay a fee to use the service for a particular time period, often used for SaaS services.

Utility Computing: Utility computing is a service provisioning model in which a service provider makes computing resources and infrastructure management available to the customer as needed, and charges them for specific usage rather than a flat rate. Like other types of on-demand computing (such as grid computing), the utility model seeks to maximize the efficient use of resources and/or minimize associated costs.

Vendor lock-in: Dependency on the particular cloud vendor and difficulty moving from one cloud vendor to another due to lack of standardized protocols, APIs, data structures (schema), and service models.

Vertical cloud: A cloud computing environment that is optimized for use in a particular industry, such as health care or financial services.

Virtual private data center: Resources grouped according to specific business objectives

Virtualization: Virtualization is the creation of a virtual version of something, such as an operating system, a server, a storage device or network resources. Operating system virtualization is the use of software to allow a piece of hardware to run multiple operating system images at the same time. The technology got its start on mainframes decades ago, allowing administrators to avoid wasting expensive processing power.

Virtual Machine: A virtual machine (VM) is an environment, usually a program or operating system, which does not physically exist but is created within another environment. In this context, a VM is called a "guest" while the environment it runs within is called a "host." Virtual machines are often created to execute an instruction set different than that of the host environment. One host environment can often run multiple VMs at once. Because VMs are separated from the physical resources they use, the host environment is often able to dynamically assign those resources among them.

V**irtual private cloud (VPC)**: similar to VPN but applied to cloud computing. Can be used to bridge private cloud and public cloud environments.

APPENDICES

Appendix 1. Cloud Computing Statement by Dr. David McClure

Dr. David McClure, Associate Administrator, Office of Citizen Services and Innovative Technologies, General Services Administration, before the House Committee on Oversight and Government Reform, Subcommittee on Government Management, Organization, and Procurement.

July 1, 2010

Cloud computing enables convenient, rapid, and on-demand computer network access—most often via the internet--to a shared pool of configurable computing resources (in the form of servers, networks, storage, applications, and services). Quite simply, it is the way computing services are delivered that is revolutionary. Cloud computing allows users to provision computing capabilities rapidly and as needed; that is, to scale out and scale back as required, and to pay only for services used. Users can provision software and infrastructure cloud services on demand with minimal, if any, human intervention. Because cloud computing is based on resource pooling and broad network access there is a natural economy of scale that can result in lower costs to agencies. In addition, cloud computing offers a varied menu of service models from a private cloud operated solely for one organization to a public cloud that is available to a large industry group and the general public and owned by an organization that is selling cloud computing services.

At GSA, we think the adoption of safe and secure cloud computing by the Federal government presents an opportunity to close the IT performance gap. Various forms of cloud computing solutions are already being used in the federal government today to save money and improve services. Let me illustrate with just a few examples:

The Department of the Army Experience Center in Philadelphia is piloting the use of a customer relationship management (CRM) tool. The Center is a recruiting center that reaches out to young people who are interested in joining our armed forces. The Center wants to move to real-time recruiting and to use tools and techniques that are familiar and appeal to its young demographic. They are using a CRM provided by SalesForce to track recruits as they work with the Center. Since the tool integrates directly with e-mail, Twitter and Facebook, recruiters can maintain connections with potential candidates directly after they leave the Center. The Army estimated that to implement a traditional CRM would have cost $500,000. The cloud-based solution has been implemented at the cost of $54,000.

The Department of Energy is evaluating the cost and efficiencies resulting from leveraging cloud computing solution across the enterprise to support business and scientific services. The Lawrence Berkeley Lab has deployed over 5,000 mailboxes on Google Federal Premiere Apps and they are now evaluating the use of Amazon Elastic Compute Cloud (EC2) to handle excess capacity for computers during peak demand. The Lab estimates that they will save $1.5 million over the next five years in hardware, software and labor costs from the deployments they have made.

Finally, my own agency – GSA has moved the primary information portal, USA.gov, to a cloud-based host. This enabled the site to deliver a consistent level of access to information as new data bases are added, as peak usage periods are encountered, and as the site evolves to encompass more services. By moving to a cloud, GSA was able to reduce site upgrade time from nine months to one day; monthly downtime improved from two hours to 99.9% availability; and GSA realized savings of $1.7M in hosting services.

In addition to improved services, GSA anticipates that cloud computing will be a major factor in reducing the environmental impact of technology and help achieve important sustainability goals. Effective use of cloud computing can be part of an overall strategy to reduce the need for multiple data centers and the energy they consume. Currently, GSA is supporting OMB in working with agencies to develop plans to consolidate their data centers. Using the right deployment model – private cloud, community cloud, public cloud, or a hybrid model – can help agencies buy improved services at a lower cost within acceptable risk levels, without having to maintain expensive, separate, independent and often needlessly redundant brick and mortar data centers.

In February 2010, the Federal CIO announced the Federal Data Center Consolidation Initiative. In it, he designated two federal agency CIOs -- Richard Spires (DHS) and Michael Duffy (Treasury) – to lead the effort inside the Federal CIO Council. It also highlighted the following goals:

- ☐ Reduce the cost of data center hardware, software and operations
- ☐ Increase the overall IT security posture of the government
- ☐ Shift IT investments to more efficient computing platforms and technologies
- ☐ Promote the use of Green IT by reducing the overall energy and real estate footprint of government data centers

GSA has a significant leadership role in supporting the adoption of cloud computing in the federal government. We have concentrated our efforts on facilitating easy access to cloud-based solutions from commercial providers that meet federal requirements, enhancing agencies' capacity to analyze viable cloud computing options that meet their business and technology modernization

needs, and addressing obstacles to safe and secure cloud computing. In particular, GSA facilitates innovative cloud computing procurement options, ensures effective cloud security and standards are in place, and identifies potential multi-agency or government-wide uses of cloud computing solutions. GSA is also the information "hub" for cloud use case examples, decisional and implementation best practices, and sharing exposed risks and lessons learned. We have set up the Info.Apps.Gov site as an evolving knowledge repository for all government agencies to use and contribute their expertise.

Let me briefly highlight how GSA is specifically providing execution capabilities to empower sensible cloud computing adoption in the federal government.

Federal Cloud Computing Project Management Office

In March of 2009, the Federal Chief Information Officer (CIO) Council identified cloud computing as a priority for meeting the growing need for effective and efficient use of information technology to meet the performance and mission needs of the government. To assist in fostering cloud computing adoption, the Federal Cloud Computing Program Management Office (PMO) was created in April of 2009 at GSA. The PMO resides in the Office of Citizen Service and Innovation Technologies and is directed by Ms. Katie Lewin who directly reports to the Deputy Administrator for Innovative Technology, Mr. Sonny Bhagowalia. The Director of the PMO also meets weekly with the Federal CIO to report on progress, discuss risks and mitigations, identify promising cloud projects across the government and refine direction. The PMO also reports on its activities and results to the CIO Council Cloud Computing Executive Steering Committee (ESC). The ESC provides oversight for the Federal Cloud Computing Initiative and fosters communications among agencies on cloud computing. ESC Membership includes senior IT executives from across the entire Federal government.

The PMO provides technical and administrative leadership to cloud computing initiatives. PMO staff is drawn from GSA technical experts with some additional contractor support. The primary focus of the PMO is on the following activities:

- ☐ Support for the design and operation of the Apps.Gov cloud computing storefront and related cloud procurement initiatives
- ☐ Facilitating identification of key cloud security requirements (certification, accreditation, and authorization), particularly on a government-wide basis through a new FedRAMP initiative
- ☐ Promotion of current and planned cloud projects across the government
- ☐ Data center consolidation analysis, planning, and strategy support
- ☐ Development and open dissemination of relevant cloud computing information.

To augment their skill base, the PMO has formed working groups to address specific areas including security, standards and specific cloud-based solutions with government or multi-agency use, such as cloud-based e-mail services. The working groups are composed of staff from across the government who bring expertise and interest to address specific obstacles or define paths to adoption. Each group is chaired by a government expert. The National Institute of Standards and Technology (NIST) led both the security and the standards groups. The e-mail group is chaired by an expert from Department of the Interior.

Cloud Computing PMO Operations

Cloud Procurement

Cloud services are usually offered and purchased as commodities. This is a new way of buying IT services and requires careful research on both government requirements and industry capability to meet demand. To assist agencies in buying new commercially provided cloud services, GSA established a website -- Apps.Gov -- modeled on other GSA product and service acquisition "storefronts." The purpose of Apps.Gov is to provide easy, simple ways to find, research, and procure commercial cloud products and services. Agencies can search for software as a service (SaaS) product categorized under 33 business purpose headings and get product descriptions, price quotes, and links to more information on specific products. Usage patterns to date indicate that agencies use this information to either directly buy SaaS products or, alternatively, as a source of marketplace research that is used to support cloud procurements using other vehicles such as GSA Schedule or GSA Advantage.

Apps.Gov also has information on no-cost social media applications that have agreed to "government-friendly" Terms of Service. When a user hits the SEND REQUEST button, they are linked to their agency's social media coordinator to complete the request for use of the tool in compliance with their agency's social media policy.

To support access to cloud-based Infrastructure as a Service (IaaS), the Cloud PMO works with the Federal Acquisition Service (FAS) at GSA. FAS has primary responsibility for operating on-line acquisition services that are available for government-wide use. In May 2009, the PMO issued a Request for Information (RFI) asking the marketplace how they would address cloud computing business models, pricing, service level agreements, operational support, data management, security and standards. The responses to this RFI were incorporated into a Request for Quote (RFQ) for Infrastructure as a Service capabilities and pricing. The result will be a multiple award blanket purchase agreement that agencies can use to procure cloud-based web hosting, virtual machine, and storage services within a moderate security environment as defined by the Federal Information Security Act (FISMA). That RFQ closed yesterday and is currently in an evaluation stage.

Cloud Computing Security

One of the most significant obstacles to the adoption of cloud computing is security. Agencies are concerned about the risks of housing data off-site in a cloud if FISMA security controls and accountabilities are not in place. In other words, agencies need to have valid certification and accreditation (C&A) process and a signed Authority to Operate (ATO) in place for each cloud-based product they use. While vendors are willing to meet security requirements, they would prefer not to go through the expense and effort of obtaining a C&A and ATO for each use of that product in all the federal departments and agencies. The PMO formed a security working group, initially chaired by NIST to address this problem. The group developed a process and corresponding security controls that were agreed to by multiple agencies – which we have termed as the Federal Risk and Authorization Management Program (FedRAMP).

FedRAMP is a government-wide initiative to provide joint authorizations and continuous security monitoring services for all federal agencies with an initial focus on cloud computing. By providing a unified government-wide risk management for enterprise level IT systems, FedRAMP will enable agencies to either use or leverage authorizations with:

- ☐ Vetted interagency approach;
- ☐ Consistent application of Federal security requirements;
- ☐ Improved community-wide risk management posture; and
- ☐ Increased effectiveness and management cost savings.

FedRAMP allows agencies to use or leverage authorizations. Under this program, agencies will be able to rely upon and review security details, leverage the existing authorization, and secure agency usage of system. This should greatly reduce cost, enable rapid acquisition, and reduce effort.

FedRAMP has three components:

- ☐ Security Requirement Authorities which create government-wide baseline security requirements that are interagency developed and approved. This will initially be the Federal Cloud Computing Initiative and ultimately live with the ISIMC Working Group.
- ☐ The FedRAMP Office which will coordinate authorization packages, manage authorized system list, and provide continuous monitoring oversight. This will be managed by GSA.
- ☐ A Joint Authorization Board which will perform authorizations and on-going risk determinations to be leveraged government-wide. The board will consist of representatives from GSA, DoD, DHS and the sponsoring agency of the authorized system.

GSA is working with OMB, security groups including the Federal CIO Council's Information Security and Identity Management Committee, and the marketplace to vet this program and ensure

that it will meet the security requirements of the government while streamlining the process for industry.

Cloud Computing and Open Government

In the past decade, vast increases in the ubiquity and availability of storage space, bandwidth, and computing power have enabled a new class of internet-based applications—broadly called "web 2.0"—that focus less on one-way delivery of information and more on enabling large, diverse communities to come together, share their wisdom, and take action. Increasingly, citizens—government's customers—simply expect to find the information they want and need through the use of the on-line social networks and platforms they are rapidly adopting and use as part of their everyday lives.

As our Administrator, Martha Johnson, noted upon being sworn in February 2010: Hoarding and hiding information prevents citizens and civil servants from understanding and participating in the public process effectively…We at GSA can help change that. We can make the information more available, as a first step. And we can do much more. We can, and will, take advantage of emerging technologies for sorting, sharing, networking, collective intelligence, and using that information. Our goal is nothing short of a nation that relies not on select data and statistical boxing matches, but on accurate evidence that supports knowledge and wisdom.

Most of these new web 2.0 technologies and tools are available as cloud-based SaaS solutions and/or are hosted in cloud computing infrastructure environments. This allows the government to offer these tools and services in a very cost-efficient manner. Let me highlight a few examples:

The Common Open Government Dialogue Platform is a project undertaken by GSA in response to the Open Government Directive's mandate that agencies "incorporate a mechanism for the public to provide input on the agency's Open Government Plan." Over the course of six weeks, GSA provided interested agencies with a no-cost, law- and policy-compliant, public-facing online engagement tool, as well as training and technical support to enable them to immediately begin collecting public and employee input on their forthcoming open government plans. Since then, GSA has worked to transfer ownership of the open government public engagement tool, powered by a cloud SaaS platform called IdeaScale, to interested agencies, in a manner that provided both policy and legal compliance, as well as support for sustained engagement. The tool was launched in February 2010 across 22 federal agencies and the White House Office of Science and Technology Policy; overall resource investment was less than $10,000 – far less than the hundreds of thousands or millions of dollars that would have resulted from agencies independently pursuing and procuring IT solutions. The agencies' dialogue sites garnered over 2,100 ideas, over 3,400 comments, and over 21,000 votes during a six-week "live" period and the tool continues to be used by several agencies for a variety of other open government purposes.

USASpending.gov is a source for information collected from agencies in accordance with the Federal Funding Accountability and Transparency Act of 2006. This public facing web site is a cornerstone of the Administration's efforts to make government open and transparent. Using USAspending.gov, the public can determine how their tax dollars are spent and gain insight into the Federal spending processes across agencies. It also houses the Federal IT Dashboard, which displays details on the nearly 800 major federal IT investments based on data reported to the Office of Management and Budget. This data is also now housed in a cloud infrastructure environment maintained by NASA.

Data.gov is the central portal for citizens to find, download, and assess government data. It now hosts over 270,000 data sets covering topics ranging from healthcare to commerce to education. Data.gov was one of the first public facing government websites to deploy cloud computing successfully in government. It empowers citizens by allowing them to create personalized mash-ups of information from diverse sources (e.g., local school academic scores arrayed by education spending levels), solve problems (e.g., FAA flight time arrival information), and build awareness of government's role in activities affecting daily activities (e.g., food safety, weather, and the like).

Challenge.gov is a government-wide challenge platform that will be hosted in a cloud computing infrastructure service to facilitate government innovation through challenges and prizes. This tool provides forums for seekers (the federal agency challenger looking for solutions) and solvers (those with potential solutions) to suggest, collaborate on, and deliver solutions. It will also allow the public to easily find and interact with federal government challenges. The platform responds to requirements defined in a March 8, 2010, OMB Memo, "Guidance on the Use of Challenges and Prizes to Promote Open Government" which included a requirement to provide a web-based challenge platform within 120 days. GSA is also exploring acquisition options to make it easier for agencies to procure products and services related to challenges.

Citizen Engagement Platform will provide a variety of blog, challenge and other engagement tools to make it easy for government to engage with citizens, and easy for citizens to engage with government. The platform addresses agencies' need for easy-to-use, easy-to-deploy, secure and policy-compliant tools. This "build once, use many" approach adds lightweight, no-cost options for agencies to create a more open, transparent and collaborative government with tools either hosted or directly managed by GSA.

Conclusion

Mr. Chairman, cloud computing has a promising future in transforming the federal government because of its ability to fundamentally reshape government IT operations used for critical government business process and citizen service delivery support. It can help shift our

focus to value added use of the information we collect and provide cost effective services in a digitally and networked enabled world. Additionally, it has the potential to free up resources that have gone to support data centers and capabilities that are better leveraged across the community – at bureau, agency or cross-agency level. At GSA, we are supporting this transformation by leveraging cloud solutions and acquisitions on a government-wide basis wherever possible to maximize economies of scale.

Thank you for the opportunity to appear today and I look forward to answering questions from you and members.

Appendix 2. FedRAMP

FedRAMP Introduction

The Federal Risk and Authorization Management Program or FedRAMP has been established to provide a standard approach to Assessing and Authorizing (A&A) cloud computing services and products. FedRAMP allows joint authorizations and continuous security monitoring services for Government and Commercial cloud computing systems intended for multi-agency use. Joint authorization of cloud providers results in a common security risk model that can be leveraged across the Federal Government. The use of this common security risk model provides a consistent baseline for cloud-based technologies. This common baseline ensures that the benefits of cloud-based technologies are effectively integrated across the various cloud computing solutions currently proposed within the government. The risk model will also enable the government to "approve once, and use often" by ensuring multiple agencies gain the benefit and insight of the FedRAMP's Authorization and access to service provider's authorization packages.

Proposed Security Assessment and Authorization for U.S. Government Cloud Computing:

- ☐ Over the past 18 months, an inter-agency team comprised of the National Institute of Standards and Technology (NIST), General Services Administration (GSA), the CIO Council and working bodies such as the Information Security and Identity Management Committee (ISIMC), has worked on developing the Proposed Security Assessment and Authorization for U.S. Government Cloud Computing.

The attached document is a product of 18 months of collaboration with State and Local Governments, Private Sector, NGO's and Academia. This marks an early step toward our goal of deploying secure cloud computing services to improve performance and lower the cost of government operations.

Executive Overview

The ability to embrace cloud computing capabilities for federal departments and agencies brings advantages and opportunities for increased efficiencies, cost savings, and green computing

technologies. However, cloud computing also brings new risks and challenges to securely use cloud computing capabilities as good stewards of government data. In order to address these concerns, the U.S. Chief Information Officer (U.S. CIO) requested the Federal CIO Council launch a government-wide risk and authorization management program. This document describes a government-wide Federal Risk and Authorization Management Program (FedRAMP) to provide joint security assessment, authorizations and continuous monitoring of cloud computing services for all Federal Agencies to leverage.

Cloud computing is not a single capability, but a collection of essential characteristics that are manifested through various types of technology deployment and service models. A wide range of technologies fall under the title "cloud computing," and the complexity of their various implementations may result in confusion among program managers. The guidelines embraced in this document, represent a subset of the National Institute of Standards and Technology (NIST) definition of cloud computing, with three service models; Software as a Service, Platform as a Service, and Infrastructure as a Service (SaaS, PaaS, and IaaS).

The decision to embrace cloud computing technology is a risk-based decision, not a technology-based decision. As such, this decision from a risk management perspective requires inputs from all stakeholders, including the CIO, CISO, Office of General Counsel (OGC), privacy official and the program owner. Once the business decision has been made to move towards a cloud computing environment, agencies must then determine the appropriate manner for their security assessments and authorizations.

Cloud Computing and Government-Wide Risk And Authorization

Cloud Computing systems are hosted on large, multi-tenant infrastructures. This shared infrastructure provides the same boundaries and security protocols for each customer. In such an environment, completing the security assessment and authorization process separately by each customer is redundant. Instead, a government-wide risk and authorization program would enable providers and the program office to complete the security assessment and authorization process once and share the results with customer agencies.

Additionally, the Federal Information Security Management Act (FISMA) and NIST special publications provide Federal Agencies with guidance and framework needed to securely use cloud systems. However, interpretation and application of FISMA requirements and NIST Standards vary greatly from agency to agency. Not only do agencies have varying numbers of security requirements at or above the NIST baseline, many times additional requirements from multiple agencies are not compatible on the same system. A government-wide risk and authorization program for cloud computing would allow agencies to completely leverage the work of an already completed authorization or only require an agency to complete delta requirements (i.e. unique requirements for that individual agency). Finally, security authorizations have become increasingly time-consuming and costly both for the Federal Government and private industry. As depicted in Figure 25, government-wide risk and authorization program will promote faster and cost effective acquisition of cloud computing systems by using an 'authorize once, use many' approach to leveraging security authorizations. Additionally, such a program will promote the Administration's goal of openness and transparency in government. All of the security requirements, processes, and templates will have to be made publicly available for consumption not only by Federal agencies but private vendors as well. This will allow Federal Agencies to leverage this work at their agency but private industry will also finally have the full picture of what a security authorization will entail prior to being in a contractual relationship with an agency.

Standardized Assessment and Authorization: FedRAMP

The Federal Risk and Authorization Management Program (FedRAMP) is designed to solve the security authorization problems highlighted by cloud computing. FedRAMP will provide a unified government-wide risk management process for cloud computing systems. FedRAMP will work in an open and transparent manner with Federal Agencies and private industry about the Assessment and Authorization process.

Through this government-wide approach, FedRAMP will enable agencies to either use or leverage authorizations with an:

- Interagency vetted approach using common security requirements;
- Consistent application of Federal security requirements;
- Consolidated risk management; and
- Increased effectiveness and management cost savings.

In addition, FedRAMP will work in collaboration with the CIO Council and Information Security and Identity Management Committee (ISIMC) to constantly refine and keep this document up to date with cloud computing security best practices. Separate from FedRAMP, ISIMC has developed guidance for agency use on the secure use of cloud computing in Federal Security Guidelines for Cloud Computing.

Transparent Path For Secure Adoption Of Cloud Computing

The security guidance and FedRAMP assessment and authorization process aims to develop robust cloud security governance for the Federal Government. The collective work that follows represents collaboration amongst security experts and representatives throughout government including all of the CIO Council Agencies.

By following the requirements and processes in this document, Federal agencies will be able to take advantage of cloud-based solutions to provide more efficient and secure IT solutions when delivering products and services to its customers.

Figure 25. FedRAMP eliminates redundancy

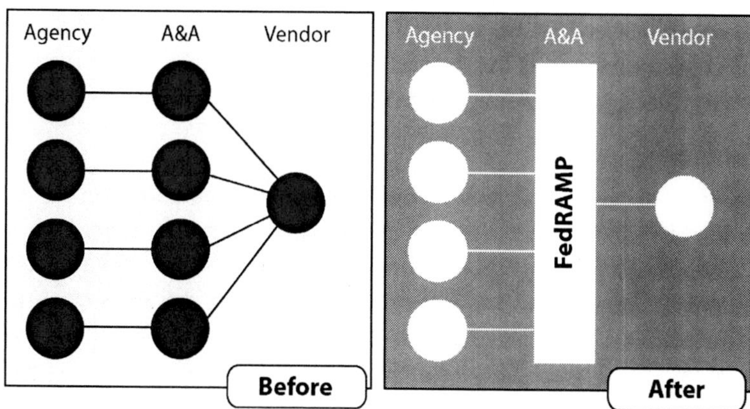

Cloud Computing Security Requirements Baseline

In the case of FedRAMP, two sets of security controls have been defined for low-impact and moderate-impact cloud information systems respectively. The impact levels are based on the sensitivity and criticality of the federal information being processed, stored, and transmitted by cloud information systems as defined in Federal Information Processing Standard 199. All NIST security standards and guidelines used to define the requirements for the FedRAMP cloud computing initiative are publicly available at http://csrc.nist.gov.

The FedRAMP defined security controls are organized by the 17 control families identified in NIST Special Publication 800-53, Revision 3 and provide the following information:

- ☐ Control Number and Name – The control number and control name relate to the control as defined in NIST Special Publication 800-53, Revision 3.
- ☐ Control Baseline – The control is listed in either the Low or Moderate impact column where applicable to that baseline. If the control is not applicable, a blank will appear in that column. If a control enhancement is applicable, the enhancement is designated inside of parenthesis. Additional security controls and control enhancements that are not included in the low and moderate control baselines defined in NIST Special Publication 800-53 Revision 3 (Appendix D) are denoted in bold font. For example, AC-2 : Control is included in the NIST Baseline AC-2 (1) : Control enhancement is included in the NIST Baseline AC-2 (7) : FedRAMP specific control enhancement.

- ☐ Control Parameter Requirements – Certain controls are defined with implementation parameters. These parameters identify the scope, frequency and other considerations for how cloud service providers address specific controls and enhancements.
- ☐ Additional Requirements and Guidance – These entries represent additional required security controls for cloud computing applications and environments of operation selected from the security control catalog in NIST Special Publication 800-53 Revision 3 (Appendix F). Required parameter values for the variable parts of security controls and control enhancements (designated by assignment and selection statements) are also provided.

Continuous Monitoring

A critical aspect of managing risk to information from the operation and use of information systems involves the continuous monitoring of the security controls employed within or inherited by the system. Conducting a thorough point-in-time assessment of the deployed security controls is a necessary but not sufficient condition to demonstrate security due diligence. An effective organizational information security program also includes a rigorous continuous monitoring program integrated into the System Development Life Cycle (SDLC). The objective of the continuous monitoring program is to determine if the set of deployed security controls continue to be effective over time in light of the inevitable changes that occur. Continuous monitoring is a proven technique to address the security impacts on an information system resulting from changes to the hardware, software, firmware, or operational environment. A well-designed and well-managed continuous monitoring program can effectively transform an otherwise static security control assessment and risk determination process into a dynamic process that provides essential, near real-time security status-related information to organizational officials in order to take appropriate risk mitigation actions and make cost-effective, risk-based decisions regarding the operation of the information system. Continuous monitoring programs provide organizations with an effective mechanism to update Security Plans, Security Assessment Reports, and Plans of Action and Milestones (POA&Ms).

An effective continuous monitoring program includes:

- ☐ Configuration management and control processes for information systems;
- ☐ Security impact analyses on proposed or actual changes to information systems and environments of operation;
- ☐ Assessment of selected security controls (including system-specific, hybrid, and common controls) based on the defined continuous monitoring strategy;
- ☐ Security status reporting to appropriate officials; and
- ☐ Active involvement by authorizing officials in the ongoing management of information system-related security risks.

Purpose

The purpose is to establish and define how Continuous Monitoring will work in a cloud computing environment and specifically within the FedRAMP framework. This document will also serve to define reporting responsibilities and frequency for the Cloud Service Offering Service Provider (CSP).

Background

Service Provider is required to develop a strategy and implement a program for the continuous monitoring of security control effectiveness including the potential need to change or supplement the control set, taking into account any proposed/actual changes to the information system or its environment of operation. Continuous monitoring is integrated into the organization's system development life cycle processes. Robust continuous monitoring requires the active involvement of information system owners and common control providers, chief information officers, senior information security officers, and authorizing officials. Continuous monitoring allows an organization to:

(i) track the security state of an information system on a continuous basis; and

(ii) maintain the security authorization for the system over time in highly dynamic environments of operation with changing threats, vulnerabilities, technologies, and missions/ business processes. Continuous monitoring of security controls using automated support tools facilitates near real-time risk management and represents a significant change in the way security authorization activities have been employed in the past. Near real-time risk management of information systems can be accomplished by employing automated support tools to execute various steps in the Risk Management Framework including authorization-related activities. In addition to vulnerability scanning tools, system and network monitoring tools, and other automated support tools that can help to determine the security state of an information system, organizations can employ automated security management and reporting tools to update key documents in the authorization package including the security plan, security assessment report, and plan of action and milestones. The documents in the authorization package are considered "living documents" and updated accordingly based on actual events that may affect the security state of the information system.

Continuous Monitoring Requirements

FedRAMP is designed to facilitate a more streamlined approach and methodology to continuous monitoring. Accordingly, service providers must demonstrate their ability to perform routine tasks on a specifically defined scheduled basis to monitor the cyber security posture of the defined IT security boundary. While FedRAMP will not prescribe specific toolsets to perform these functions, FedRAMP does prescribe their minimum capabilities. Furthermore, FedRAMP

will prescribe specific reporting criteria that service providers can utilize to maximize their FISMA reporting responsibilities while minimizing the resource strain that is often experienced.

Reporting and Continuous Monitoring

Maintenance of the security Authority To Operate (ATO) will be through continuous monitoring of security controls of the service providers system and its environment of operation to determine if the security controls in the information system continue to be effective over time in light of changes that occur in the system and environment. Through continuous monitoring, security controls and supporting deliverables are updated and submitted to FedRAMP per the schedules below. The submitted deliverables provide a current understanding of the security state and risk posture of the information systems. They allow FedRAMP authorizing officials to make credible risk-based decisions regarding the continued operations of the information systems and initiate appropriate responses as needed when changes occur. The deliverable frequencies below are to be considered standards. However, there will be instances, beyond the control of FedRAMP in which deliverables may be required on an ad hoc basis. The deliverables required during continuous monitoring are depicted the table: FedRAMP Continuous Monitoring . This table provides a listing of the deliverables, responsible party and frequency for completion. The table is organized into:

- ☐ Deliverable – Detailed description of the reporting artifact. If the artifact is expected in a specific format, that format appears in **BOLD** text.
- ☐ Frequency – Frequency under which the artifact should be created and/or updated.
- ☐ Responsibility – Whether FedRAMP or the Cloud Service Provider is responsible for creation and maintenance of the artifact.

Deliverable Table

Deliverable	Frequency	Responsibility	
		FedRAMP	Cloud Service Provider
Scan reports of all systems within the boundary for vulnerability (Patch) management **(Tool Output Report)**	Monthly		✓
Scan for verification of FDCC compliance (USGCB, CIS) **(SCAP Tool Output)**	Quarterly		✓
Incident Response Plan	Annually		✓
POAM Remediation **(Completed POA&M Matrix)**	Quarterly		✓
Change Control Process	Annually		✓

Deliverable	Frequency	Responsibility	
		FedRAMP	Cloud Service Provider
Penetration testing **(Formal plan and results)**	Annually	✓	✓
IV&V of controls	Semi-Annually	✓	✓
Scan to verify that boundary has not changed (also that no rogue systems are added after ATO) **(Tool Output Report)**	Quarterly		✓
FISMA Reporting Data	Quarterly		✓
Update Documentation	Quarterly		✓
Contingency Plan and Test Report	Annually		✓
Separation of Duties Matrix	Annually		✓
Information Security Awareness and Training Records Results	Annually		✓

Routine Systems Change Control Process

The Change Control Process is instrumental in ensuring the integrity of the cloud computing environment. As the system owners as well as other authorizing officials approve changes, they are systematically documented. This documentation is a critical aspect of continuous monitoring since it establishes all of the requirements that led to the need for the change as well as the specific details of the implementation. To ensure that changes to the enterprise do not alter the security posture beyond the parameters set by the FedRAMP Joint Authorization Board (JAB) the key documents in the authorization package which include the security plan, security assessment report, and plan of action and milestones are updated and formally submitted to FedRAMP within 30 days of approved modification.

There are however, changes that are considered to be routine. These changes can be standard maintenance, addition or deletion of users, the application of standard security patches, or other routine activities. While these changes individually may not have much effect on the overall security posture of the system, in aggregate they can create a formidable security issue. To combat this possibility, these routine changes should be documented as part of the CSP's standard change management process and accounted for via the CSP's internal continuous monitoring plan. Accordingly, these changes must be documented, at a minimum, within the current SSP of the system within 30 days of implementation.

Configuration Change Control Process (CCP)

Throughout the System Development Life Cycle (SDLC) system owners must be cognizant of changes to the system. Since systems routinely experience changes over time to accommodate new requirements, new technologies or new risks, they must be routinely analyzed in respect to the security posture. Minor changes typically have little impact to the security posture of a system. These changes can be standard maintenance, adding or deleting users, applying standard security patches, or other routine activities. However, significant changes require an added level of attention and action. NIST defines significant change as "A significant change is defined as a change that is likely to affect the security state of an information system." Changes such as installing a new operating system, port modification, new hardware platforms, or changes to the security controls should automatically trigger a re-authorization of the system via the FedRAMP process.

Minor changes must be captured and documented in the SSP of the system within 30 days of implementation. This requirement should be part of the CSP's documented internal continuous monitoring plan. Once the SSP is updated, it must be submitted to FedRAMP, and a record of the change must be maintained internally. Major or significant changes may require re-authorization via the FedRAMP process. In order to facilitate a re-authorization, it is the responsibility of both the CSP and the sponsoring agency to notify FedRAMP of the need to make such a significant change. FedRAMP will assist and coordinate with all stakeholders the necessary steps to ensure that the change is adequately documented, tested and approved.

FISMA Reporting Requirements

FISMA established the IT security reporting requirements. OMB in conjunction with DHS enforces these reporting requirements. FISMA reporting responsibilities must be clearly defined.

FedRAMP will coordinate with CSP's and agencies to gather data associated with the cloud service offering. Only data related to the documented system security boundary of the cloud service offering will be collected by FedRAMP and reported to OMB at the appropriate time and frequency. Agencies will maintain their reporting responsibilities for their internal systems that correspond to the inter-connection between the agency and the cloud service offering.

On-going Testing of Controls and Changes to Security Controls

Process

System owners and administrators have long maintained the responsibility for patch and vulnerability management. However, it has been proven time and again that this responsibility often requires a heavy use of resources as well as a documented, repeatable process to be carried out consistently and adequately. This strain on resources and lack of processes has opened the door to many malicious entities through improper patching, significant lapse in time between patch availability and patch implementation, and other security oversights. Routine system scanning and

reporting is a vital aspect of continuous monitoring and thus, maintaining a robust cyber security posture.

Vulnerability patching is critical. Proprietary operating system vendors (POSV) are constantly providing patches to mitigate vulnerabilities that are discovered. In fact, regularly scheduled monthly patches are published by many POSV to be applied to the appropriate operating system. It is also the case that POSV will, from time to time, publish security patches that should be applied on systems as soon as possible due to the serious nature of the vulnerability. Systems running in virtual environment are not exempted from patching. In fact, not only are the operating systems running in a virtual environment to be patched routinely, but often-times the virtualization software itself is exposed to vulnerabilities and thus must be patched either via a vendor-based solution or other technical solution.

Open source operating systems require patch and vulnerability management as well. Due to the open nature of these operating systems there needs to be a reliable distribution point for system administrators to safely and securely obtain the required patches. These patches are available at the specific vendors' website.

Database platforms, web platforms and applications, and virtually all other software applications come with their own security issues. It is not only prudent, but also necessary to stay abreast of all of the vulnerabilities that are represented by the IT infrastructure and applications that are in use.

While vulnerability management is indeed a difficult and daunting task, there are proven tools available to assist the system owner and administrator in discovering the vulnerabilities in a timely fashion. These tools must be updated prior to being run. Updates are available at the corresponding vendors' website.

With these issues in mind FedRAMP will require CSP's to provide the following:

- ☐ Monthly vulnerability scans of all servers. Tools used to perform the scan must be provided as well as the version number reflecting the latest update. A formal report of all vulnerabilities discovered, mitigated or the mitigating strategy. This report should list the vulnerabilities by severity and name. Specificity is crucial to addressing the security posture of the system. All "High" level vulnerabilities must be mitigated within thirty days (30) days of discovery. "Moderate" level vulnerabilities must be mitigated within ninety (90) days of discovery. It is accepted that, at certain times, the application of certain security patches can cause negative effects on systems. In these situations, it is understood that compensating controls (workarounds) must be used to minimize system performance degradation while serving to mitigate the vulnerability. These "Workarounds" must be submitted to FedRAMP and the Sponsoring agency for acceptance. All reporting must reflect these activities.

- ☐ Quarterly FDCC and/or system configuration compliance scans, with a Security Content Automation Protocol (SCAP) validated tool, across the entire boundary, which verifies that all servers maintain compliance with the mandated FDCC and/or approved system configuration security settings.
- ☐ Weekly scans for malicious code. Internal scans must be performed with the appropriate updated toolset. Monthly reporting is required to be submitted to FedRAMP, where activity is summarized. All software operating systems and applications are required to be scanned by an appropriate tool to perform a thorough code review to discover malicious code. Mandatory reporting to FedRAMP must include tool used, tool configuration settings, scanning parameters, application scanned (name and version) and the name of the third party performing the scan. Initial report should be included with the SSP as part of the initial authorization package.
- ☐ Performance of the annual Self-Assessment in accordance with NIST guidelines. CSP must perform a self-assessment annually or whenever a significant change occurs. This is necessary if there is to be a continuous awareness of the risk and security posture of the system.
- ☐ Quarterly POA&M remediation reporting. CSP must provide to FedRAMP a detailed matrix of POA&M activities using the supplied FedRAMP POA&M Template. This should include milestones met or milestones missed, resources required and validation parameters.
- ☐ Active Incident Response capabilities allow for suspect systems to be isolated and inspected for any unapproved or otherwise malicious applications.
- ☐ Quarterly boundary-wide scans are required to be performed on the defined boundary IT system inventory to validate the proper HW and SW configurations as well as search and discover rogue systems attached to the infrastructure. A summary report, inclusive of a detailed network architecture drawing must be provided to FedRAMP. Change Control Process meetings to determine and validate the necessity for suggested changes to HW/SW within the enterprise must be coordinated with FedRAMP to ensure that the JAB is aware of the changes being made to the system.

Incident Response

Computer security incident response has become an important component of information technology (IT) programs. Security-related threats have become not only more numerous and diverse but also more damaging and disruptive. New types of security-related incidents emerge frequently. Preventative activities based on the results of risk assessments can lower the number of incidents, but not all incidents can be prevented. An incident response capability is therefore necessary for rapidly detecting incidents, minimizing loss and destruction, mitigating the weaknesses that were exploited, and restoring computing services. To that end, NIST SP 800-61

provides guidelines for development and initiation of an incident handling program, particularly for analyzing incident-related data and determining the appropriate response to each incident.

The guidelines can be followed independently of particular hardware platforms, operating systems, protocols, or applications. As part of the authorization process the system security plan will have documented all of the "IR" or Incident Response family of controls. One of these controls (IR-8) requires the development of an Incident Response plan that will cover the life cycle of incident response as documented in the NIST SP 800-61 guidelines. The plan should outline the resources and management support that is needed to effectively maintain and mature an incident response capability. The incident response plan should include these elements:

- ☐ Mission
- ☐ Strategies and goals
- ☐ Senior management approval
- ☐ Organizational approach to incident response
- ☐ How the incident response team will communicate with the rest of the organization
- ☐ Metrics for measuring the incident response capability
- ☐ Roadmap for maturing the incident response capability
- ☐ How the program fits into the overall organization.

The organization's mission, strategies, and goals for incident response should help in determining the structure of its incident response capability. The incident response program structure should also be discussed within the plan. The response plan must address the possibility that incidents, including privacy breaches and classified spills, may impact the cloud and shared cloud customers. In any shared system, communication is the biggest key to success.

As part of the continuous monitoring of a system, responding to incidents will be a key element.

The FedRAMP concern and its role in continuous monitoring will be to focus on how a provider conducted the incident response and any after incident actions. As represented in Figure 26: Incident response life cycle, incident response is a continually improving process.

Figure 26. Incident response life cycle

Preparation → Detection and Analysis → Containment, Eradication, and Recovery → Post-Incident Activity → (back to Preparation)

One of the most important parts of incident response is also the most often omitted - learning and improving. Each incident response team should evolve to reflect new threats, improved technology, and lessons learned. Many organizations have found that holding a "lessons learned" meeting with all involved parties after a major incident, and periodically after lesser incidents, is extremely helpful in improving security measures and the incident handling process itself. This meeting provides a chance to achieve closure with respect to an incident by reviewing what occurred, what was done to intervene, and how well intervention worked. The meeting should be held within several days of the end of the incident. Questions to be answered in the lessons learned meeting include:

- ☐ Exactly what happened, and at what times?
- ☐ How well did staff and management perform in dealing with the incident? Were the documented procedures followed? Were they adequate?
- ☐ What information was needed sooner?
- ☐ Were any steps or actions taken that might have inhibited the recovery?
- ☐ What would the staff and management do differently in a future occurrence?
- ☐ What corrective actions can prevent similar incidents in the future?
- ☐ What tools/resources are needed to detect, analyze, and mitigate future incidents?

Small incidents need limited post-incident analysis, with the exception of incidents performed through new attack methods that are of widespread concern and interest. After serious attacks have occurred, it is usually worthwhile to hold post-mortem meetings that cross team and organizational boundaries to provide a mechanism for information sharing. The primary consideration in holding such meetings is ensuring that the right people are involved. Not only is it important to invite people who have been involved in the incident that is being analyzed, but also wise to consider who should be invited for the purpose of facilitating future cooperation.

Independent Verification and Validation

Independent Verification and Validation (IV&V) is going to be an integral component to a successful implementation of FedRAMP. With this in mind, it must be noted that establishing and maintaining an internal expertise of FedRAMP policies, procedures and processes is going to be required. This expertise will be tasked to perform various IV&V functions with CSPs, sponsoring agencies and commercial entities obtained by CSP's with absolute independence on behalf of FedRAMP. FedRAMP IV&V will be on behalf of the JAB.

As part of these efforts, FedRAMP will periodically perform audits (both scheduled and unscheduled) related strictly to the cloud computing service offering and the established system boundary. This will include, but not be limited to:

- ☐ Scheduled annual assessments of the system security documentation;
- ☐ Verification of testing procedures;
- ☐ Validation of testing tools and assessments;
- ☐ Validation of assessment methodologies employed by the CSP and independent assessors;
- ☐ Verification of the CSP continuous monitoring program; and
- ☐ Validation of CSP risk level determination criteria.

There are several methods that must be employed to accomplish these tasks. In accordance with the new FIMSA requirement, and as a matter of implementing industry best practices, FedRAMP IV&V will be performing penetration testing. This testing will be performed with strict adherence to the specific guidelines established by a mutually agreed upon "Rules of Engagement" agreement between FedRAMP IV&V and the target stakeholders. Unless otherwise stated in the agreement, all penetration testing will be passive in nature to avoid unintentional consequences. No attempts to exploit vulnerabilities will be allowed unless specified within the "Rules of Engagement" agreement.

Potential Assessment and Authorization Approach

Cloud computing presents a unique opportunity to increase the effectiveness and efficiency of the A&A and Continuous Monitoring process for Federal Agencies. The nature of cloud computing systems does not allow Federal Agencies to enforce their own unique security requirements and policies on a shared infrastructure – as many of these unique requirements are incompatible. Hence, cloud computing provides an opportunity for the Federal Agencies to work together to create a common security baseline for authorizing these shared systems.

The implementation of a common security baseline requires a joint approach for the A&A and Continuous Monitoring process. Any joint approach to this process requires a coordinated effort of many operational components working together. These operations need to interact/interplay with each other to successfully authorize and monitor cloud systems for government-wide use. FedRAMP operations could potentially be executed by different entities and in many different models. However, the end goal is to establish an on-going A&A approach that all Federal Agencies can leverage. To accomplish that goal, the following benefits are desired regardless of the operating approach:

- ☐ Inter-agency vetted Cloud Computing Security Requirement baseline that is used across the Federal Government;
- ☐ Consistent interpretation and application of security requirement baseline in a cloud computing environment;

- ☐ Consistent interpretation of cloud service provider authorization packages using a standard set of processes and evaluation criteria;
- ☐ More consistent and efficient continuous monitoring of cloud computing environment/systems fostering cross-agency communication in best practices and shared knowledge; and
- ☐ Cost savings/avoidance realized due to the "Approve once, use often" concept for security authorization of cloud systems.

FedRAMP operations could be conducted under many delivery models. The Federal Cloud Computing Initiative (FCCI) has focused on exploring three models in particular. The three models for assessment that have been vetted within Government and Industry are:

- ☐ A centralized approach working through a FedRAMP program office;
- ☐ A federated model using capabilities of multiple approved agency centers; and
- ☐ Some combination of the above that combines public and private sector partners.

Preliminary vetting of the three models focused on finding a model that best met the goals of this endeavor as mentioned above. As a result of vetting the models with government and industry stakeholders, this chapter presents FedRAMP operations through a centralized program office context. However, the government is seeking your input, knowledge, and experience as to the best model for FedRAMP operations that deliver upon the described benefits and encourage you to actively engage and contribute with substantive comments.

Overview

Background

The Federal Government is increasingly using large shared and outsourced systems by moving to cloud computing, virtualization, and datacenter/application consolidation. The current method of conducting risk management of shared, outsourced, cloud computing systems on an agency-by-agency basis causes duplication of efforts, inefficiencies in sharing knowledge, best practices and lessons learned in authorizing and ongoing monitoring of such systems, and the unnecessary cost from repetitive work and relearning. In order to address these concerns, the U.S. Chief Information Officer (U.S. CIO) established a government-wide Federal Risk and Authorization Management Program (FedRAMP) to provide joint security assessment, authorizations and continuous monitoring of cloud computing services for all Federal Agencies to leverage.

Purpose

The objective of FedRAMP is threefold:

- ☐ Ensure that information systems/services used government-wide have adequate information security;

- ☐ Eliminate duplication of effort and reduce risk management costs; and
- ☐ Enable rapid and cost-effective procurement of information systems/services for Federal agencies.

Benefits

Joint authorization of cloud computing services provides a common security risk model that can be leveraged across the Federal Government. The use of a common security risk model provides a consistent baseline for cloud-based technologies across government. This common baseline will ensure that the benefits and challenges of cloud-based technologies are effectively integrated across the various cloud computing solutions currently proposed within the government. The risk model will also enable the government to "approve once and use often" by ensuring other agencies gain the benefit and insight of the FedRAMP's Authorization and access to service providers' authorization packages.

By providing a unified government-wide risk management for enterprise level IT systems, FedRAMP will enable agencies to either use or leverage authorizations with:

- ☐ An interagency vetted approach;
- ☐ Consistent application of Federal security requirements;
- ☐ Consolidated risk management; and
- ☐ Increased effectiveness and management cost savings.

Governance

The following sections describe the FedRAMP governance model and define the roles and responsibilities of key stakeholders of the FedRAMP process.

Governance Model

Figure 27. FedRAMP Governance Model

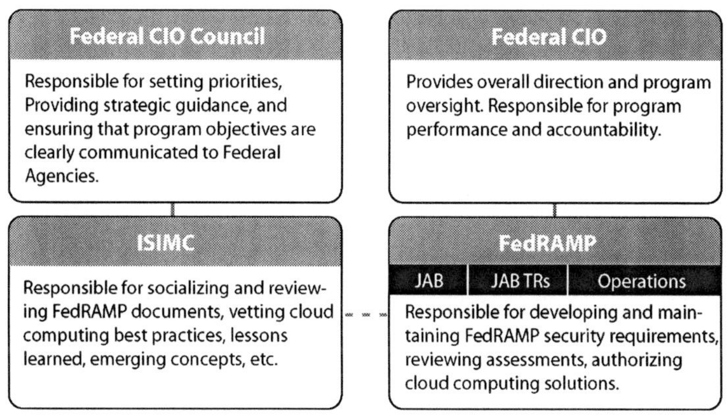

FedRAMP is an interagency effort under the authority of the U.S. Chief Information Officer (U.S. CIO) and managed out of the General Services Administration (GSA) as depicted in Figure 27 and detailed below.

The initiation of FedRAMP and the Joint Authorization Board (JAB) has been via the U.S. CIO in coordination with the Federal CIO Council. The U.S. CIO has tasked the Joint Authorization Board (JAB) with jointly authorizing cloud computing systems. The General Service Administration has been tasked with the actual day-to-day operation of FedRAMP in supporting this effort.

The three permanent members of JAB include the Department of Homeland Security (DHS), Department of Defense (DOD), and the General Services Administration (GSA). The sponsoring government agency for each cloud computing system will be represented as the rotating JAB member. The JAB also performs risk determination and acceptance of FedRAMP authorized systems.

The JAB also has the final decision making authority on FedRAMP security controls, policies, procedures and templates. JAB technical representatives are appointed by their respective JAB authorizing official (both permanent and rotating) for the implementation of the FedRAMP process. JAB technical representatives provide subject matter expertise and advice to the JAB authorizing officials.

The JAB technical representatives review the vetted authorization packages provided by FedRAMP. The JAB technical representatives make authorization recommendations to the JAB authorizing officials and advise the JAB of all residual risks. FedRAMP is an administrative support team provided by the U.S. CIO under the guidance of GSA. FedRAMP operations are responsible for the day-to-day administration and project management of FedRAMP. FedRAMP performs an initial review of submitted authorization packages and has the authority to work with cloud computing system owners to refine each submission until it satisfies FedRAMP and JAB requirements. FedRAMP also oversees continuous monitoring of authorized systems. The ISIMC under the Federal CIO Council is responsible for socializing and reviewing FedRAMP processes and documents. They provide recommendations on the FedRAMP documents directly to the JAB. Their recommendations are based on vetting the cloud computing best practices, lessons learned and emerging concepts within the Federal CIO Council community. However, the final approval on changes to FedRAMP processes and documents is made by the JAB.

Roles and Responsibilities

Table 3: Stakeholder Roles and Responsibilities defines the responsibilities/tasks for FedRAMP stakeholders

Role	Duties and Responsibilities
JAB Chair (U.S. CIO)	• Selects the JAB Authorizing Officials • Coordinates FedRAMP activities with the CIO Council • Tasks and funds FedRAMP, for technical support as necessary
JAB Authorizing Officials	• Designate a JAB Technical Representative • Ensure the Technical Representative considers current threats and evaluation criteria based on evolving cloud computing best practices in their review of joint authorizations. • Issue joint authorization decisions • Resolve issues as needed
JAB Rotating Authorizing Officials (Sponsoring Agency Authorizing Official)	• Same duties of JAB Authorizing Officials only for their sponsored cloud solution
FedRAMP Operations	• Communicate FedRAMP security requirements to service providers for prospective providers • Review CSP security authorization packages • Work with JAB Technical Representatives to clarify questions and concerns regarding authorization packages • Maintain a repository of Authorizations in two categories – Authorizations granted by the JAB. – Authorizations granted by individual agencies. • Perform continuous monitoring oversight of FedFAMP authorized systems. • Collect FISMA data from FedRAMP authorized systems for quarterly and annually reporting of data to OMB through GSA. • Facilitate the leveraging of authorized systems for other federal entities. • Maintain knowledge of the FedRAMP capabilities and process throughout industry and the federal government.
JAB Technical Representatives (including the technical representative from the sponsoring agency)	• Provide subject matter expertise to implement the direction of the JAB Authorizing Official. • Support the FedRAMP in defining and implementing the joint authorization process. • Recommend authorization decisions to the JAB Authorizing Official. • Escalate issues to the JAB Authorizing Official as appropriate.

Role	Duties and Responsibilities
Sponsoring Agency	• Cloud system selection and submission to FedRAMP • Ensures a contractual agreement with a provider is in place using FedRAMP requirements. • Designate Federal personnel to facilitate the receipt and delivery of deliverables between the cloud computing provider (CSP) and FedRAMP. • Assessment, Authorization and continuous monitoring and FISMA, reporting of controls that are agency's (customer's) responsibility.
Leveraging Agency	• Review FedRAMP authorization packages. • Determine if the stated risk determination and acceptance is consistemt with its agency's needs. • Authorize cloud system for their agency use. • Assessment, Authorization and continuous monitoring and FISMA reporting of controls that are agency's (customer's) responsibility.
Cloud Service Provider (CSP)	• The service provider is a government or commercial entity that has a cloud offering/service (IaaS, PaaS, or SaaS) and requires FedRAMP authorization of the offering/service for Government use. • Work with the sponsoring agency to submit their offering for FedRAMP authorization. • Hire independent third party assessor to perform initial system assessment and on-going monitoring of controls. • Create and submit authorization packages. • Provide Continuous Monitoring reports and updates to FedRAMP.

Assessment and Authorization Processes

High-Level Overview

The following figure depicts the high-level process for getting on the FedRAMP authorization request log. Once the Cloud Service Provider (CSP) system is officially on the FedRAMP authorization log, FedRAMP begins processing the cloud system for JAB authorization. The subsequent sections detail the steps involved in the FedRAMP Assessment and Authorization process.

Figure 28. FedRAMP authorization request process

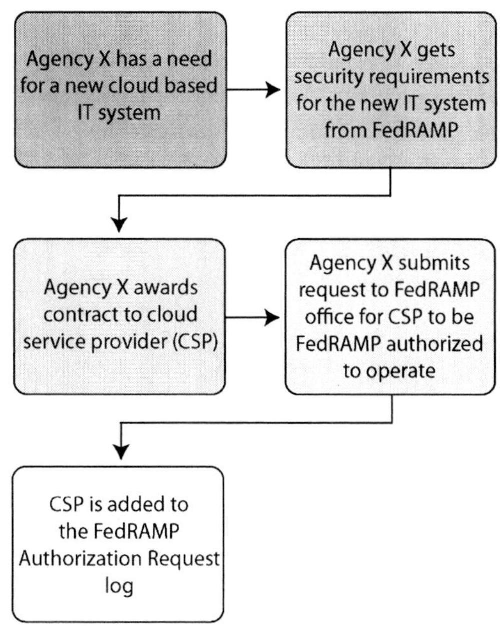

Figure 29. FedRAMP authorization process

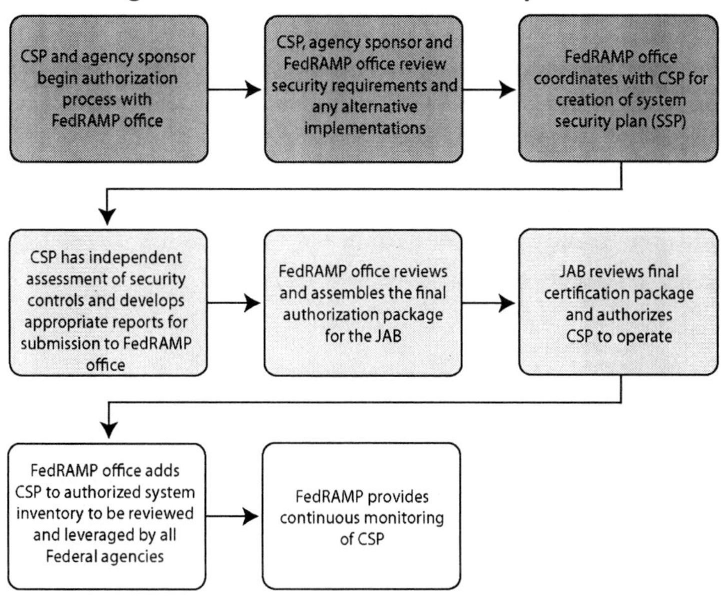

Detailed Assessment and Authorization Process

Purpose

This section defines FedRAMP assessment and authorization process for Cloud Service Providers (CSP). It also provides guidelines and procedures for applying the NIST 800-37 R1 Risk Management Framework to include conducting the activities of security categorization, security control selection and implementation, security control assessment, information system authorization, and continuous monitoring. CCS Service Providers should use this process and the noted references prior to initiating/performing the Security Authorization process.

Policy

Security Authorization Process:

a. The FedRAMP Authorizing Officials (AO) must authorize, in writing, all cloud computing systems before they go into operational service for government interest. b. A service provider's cloud computing systems must be authorized/reauthorized at least every three (3) years or whenever there is a significant change to the system's security posture in accordance with NIST SP 800-37 R1. Authorization termination dates are influenced by FedRAMP policies that may establish maximum authorization periods. For example, if the maximum authorization period for an information system is three years, then the service provider establishes a continuous monitoring strategy for assessing a subset of the security controls employed within and inherited by the system during the authorization period. This strategy allows all security controls designated in the respective security plans to be assessed at least one time by the end of the three-year period.

This also includes any common controls deployed external to service provider cloud computing systems. If the security control assessments are conducted by qualified assessors with the required degree of independence based on policies, appropriate security standards and guidelines, and the needs of the FedRAMP authorizing officials, the assessment results can be cumulatively applied to the reauthorization, thus supporting the concept of ongoing authorization. FedRAMP policies regarding ongoing authorization and formal reauthorization, if/when required, are consistent with federal directives, regulations, and/or policies.

Required Artifacts

All Service Providers' CCS must complete and deliver the following artifacts as part of the authorization process. Templates for these artifacts can be found in FedRAMP templates as described in reference materials:

- ☐ Privacy Impact Assessment (PIA)
- ☐ FedRAMP Test Procedures and Results
- ☐ Security Assessment Report (SAR)

- ☐ System Security Plan (SSP)
- ☐ IT System Contingency Plan (CP)
- ☐ IT System Contingency Plan (CP) Test Results
- ☐ Plan of Action and Milestones (POA&M)
- ☐ Continuous Monitoring Plan (CMP)
- ☐ FedRAMP Control Tailoring Workbook
- ☐ Control Implementation Summary Table
- ☐ Results of Penetration Testing
- ☐ Software Code Review
- ☐ Interconnection Agreements/Service Level Agreements/Memorandum of Agreements

Assessment and Authorization Process Workflow

FedRAMP Assessment and Authorization is an effort composed of many entities/stakeholders working together in concert to enable government-wide risk management of cloud systems.

The following provides the list of NIST special publications, FIPS publications, OMB Memorandums, FedRAMP templates and other guidelines and documents associated with the seven steps of the FedRAMP process:

Step 1 - Categorize Cloud System: (FIPS 199 / NIST Special Publications 800-30, 800-39, 800-59, 800-60.)

Step 2 – Select Security Controls: (FIPS Publications 199, 200; NIST Special Publications 800-30, 800-53 R3, FedRAMP security control baseline)

Step 3 – Authorization Request: (FedRAMP primary Authorization Request letter, FedRAMP secondary authorization request letter)

Step 4 - Implement Controls: (FedRAMP control tailoring workbook; Center for Internet Security (CIS); United States Government Configuration Baseline (USGCB); FIPS Publication 200; NIST Special Publications 800-30, 800-53 R3, 800-53A R1)

Step 5 – Assess Controls: (FedRAMP Test Procedures: Center for Internet Security (CIS); United States Government Configuration Baseline (USGCB); NIST Special Publication 800-53A R1)

Step 6 – Authorize Cloud System: OMB Memorandum 02-01; NIST Special Publications 800-30, 800-53A R1)

Step 7 – Continuous Monitoring: FedRAMP Test Procedures; NIST Special Publications 800-30, 800-53A R1, 800-37 R1

Risk Acceptability Criteria

The following table lists the FedRAMP JAB acceptable risk criteria. In particular the table lists the "Not Acceptable" risk criteria and the ones requiring JAB prior approval.

FedRAMP JAB acceptable risk criteria

Not Acceptable	Requires JAB Prior Approval
• Vulnerability Scanner output has HIGH vulnerabilities not remediated. • More than 5% of total security controls are reflected within the POA&M • False Positive claims are not supported by evidence files. • FedRAMP audit shows configuration which differs from presented documentation. • OS out of lifecycle Support (Windows XP and before). • Hot fix patches not implemented, without justification • Does not support 2-factor authentication from customer agency to cloud for moderate impact system. • Does not support FIPS 140-2 from customer agency to the cloud.	• Change in inter-connections. • Change in ISA/MOU. • Change in physical location. • Change in the Security Impact Level • Threat Change • Privacy Act security posture change. • OS Change (2k to 2k3, Windows to Linux, etc.) • Change in SW (i.e. Oracle to SQL).

Not Authorization Maintenance Process

Once a system has received a FedRAMP authorization, several events take place. First, the system is added to the FedRAMP online repository of authorized systems. Next, FedRAMP will begin facilitating agency access to the approved authorization package to enable agency review of the material. Lastly, FedRAMP will begin overseeing continuous monitoring of the system and advise the JAB of any changes to risk posture.

FedRAMP will maintain an online repository of cloud system authorizations in two categories:

☐ Authorizations granted by the JAB

☐ Authorizations granted by individual agencies

This web-based resource will be publicly accessible and will be the authoritative source of FedRAMP system authorization status. The web-based resource will maintain the following information for each currently authorized system.

- ☐ System Name and scope of authorization (examples of scope include IaaS, PaaS or SaaS, entire or partial suite of products offered by CSP)
- ☐ FIPS 199 impact level supported by the cloud system
- ☐ Expiration date for Authorization
- ☐ Version of FedRAMP requirements and templates used to authorize the system
- ☐ Points of Contact for the cloud system

FedRAMP will also maintain a secure website (separate from the public website) accessible only to Federal officials to access CSP authorization packages and communicate cloud system specific updates on the risk posture.

Authorization Leveraging Process

The purpose of all of the FedRAMP authorizations is to facilitate the leveraging of these authorizations for use by multiple federal agencies ("Approve once. Use often"). Leveraging such authorizations is employed when a federal agency chooses to accept all of the information in an existing authorization package via FedRAMP.

A FedRAMP joint authorization is not a "Federal Authority to Operate" exempting Federal Agencies, Bureaus, and Divisions from individually granting Authorities to Operate. A FedRAMP Authorization provides a baseline Authorization for Federal Agencies, Bureaus, and Divisions to review and potentially leverage. As is consistent with the traditional authorization process, an authorizing official in the leveraging organization is both responsible and accountable for accepting the security risks that may impact the leveraging organization's operations and assets, individuals, other organizations, or the Nation.

The leveraging organization reviews the FedRAMP authorization package as the basis for determining risk to the leveraging organization. When reviewing the authorization package, the leveraging organization considers risk factors such as the time elapsed since the authorization results were produced, the results of continuous monitoring, the criticality/sensitivity of the information to be processed, stored, or transmitted, as well as the overall risk tolerance of the leveraging organization.

FedRAMP will provide leveraging agencies with access to the authorization packages to assist in their risk management decision. If the leveraging organization determines that there is insufficient information in the authorization package or inadequate security measures in place for establishing an acceptable level of risk, the leveraging organization needs to communicate that to

FedRAMP. If additional information is needed or additional security measures are needed such as increasing specific security controls, conducting additional assessments, implementing other compensating controls, or establishing constraints on the use of the information system or services provided by the system these items will be facilitated by FedRAMP. The goal is to keep unique requirements to a minimum, but consider any other additional security controls for implementation and inclusion in the baseline FedRAMP security controls.

The leveraged authorization approach provides opportunities for significant cost savings and avoids a potentially costly and time-consuming authorization process by the leveraging organization. Leveraging organizations generate an authorization decision document and reference, as appropriate, information in the authorization package from FedRAMP.

All of the FedRAMP authorizations do not consider the actual information placed in the system. It is the leveraging agencies responsibility to do proper information categorization and determination if privacy information will be properly protected and if a complete Privacy Impact Assessment is in place. In almost all cases the FedRAMP authorization does not consider the actual provisioning of users and their proper security training. In all cases additional security measures will need to be documented. The leveraging organization documents those measures by creating an addendum to the original authorization package of FedRAMP or a limited version of a complete package that references the FedRAMP authorization. This addendum may include, as appropriate, updates to the security plan (for the controls that is customer agency's implementation responsibility), security assessment report, and/or leveraging organization's plan of action and milestones. FedRAMP will report the base system for FISMA purposes and the leveraging agency will need to report their authorization via their organizational FISMA process.

Consistent with the traditional authorization process, a single organizational official in a senior leadership position in the leveraging organization is both responsible and accountable for accepting the information system-related security risks that may impact the leveraging organization's operations and assets, individuals, other organizations, or the Nation.

The leveraged authorization remains in effect as long as the leveraging organization accepts the information system-related security risks and the authorization meets the requirements established by federal and/or organizational policies. This requires the sharing of information resulting from continuous monitoring activities conducted by FedRAMP and will be provided to agencies that notify FedRAMP that they are leveraging a particular package. The updates will include such items as updates to the security plan, security assessment report, plan of action and milestones, and security status reports. To enhance the security of all parties, the leveraging organization can also share with the owning organization, the results from any RMF-related activities it conducts to supplement the authorization results produced by the owning organization.

Communications Process

FedRAMP interacts with multiple stakeholders during the security lifecycle of a system. To streamline the workflow, a secure website is under development to facilitate updates on status, provide secure posting of artifacts and provide baseline information. However, in addition to this online web portal, proactive communication is required to ensure the success of each individual cloud system authorization. It is expected that the Cloud Service Providers, FedRAMP and Sponsoring and Leveraging Agencies will communicate regularly to ensure that information is disseminated effectively.

The following communication templates will be employed:

- ☐ Sponsorship Letter
- ☐ Status Report
- ☐ Confirmation Receipts (Complete Package, Incomplete Package)
- ☐ Review Recommendation (Acceptable, Unacceptable)
- ☐ Missing Artifact List
- ☐ Incident Report

The communication plan identifies the touch points and how communication will be delivered between FedRAMP, Leveraging Agencies, Sponsoring Agencies, and the Cloud Service Providers. Additional e-mails, conference calls and in-person meetings to facilitate the process as the team deems necessary may augment the communication plan. As changes are integrated into the requirement process, the communication plan may be updated to respond to required changes to the communication process. At a minimum, the communication plan will be reviewed annually. The plan is organized by phases and depicts the communication flow in the following areas:

- ☐ Trigger Event – Identifies the event that will start the require communication during the different operational processes of FedRAMP
- ☐ Deliverable –Artifact used to communicate the results/output of the trigger event to FedRAMP stakeholders
- ☐ Initiator – The entity responsible for starting the communication process.
- ☐ Target Audience – Receivers of the deliverable in the communication process.
- ☐ Delivery Method – How the artifacts will be communicated to the target audience.

Change Management Process

The technology changes within the dynamic and scalable cloud computing environment are expected to be quite swift. As the cloud computing market matures, best practices associated with the implementation and testing of security controls will evolve. There are multiple industry groups, academic collaborations, engineering teams, policy firms and assorted cadre of experts striving to

maximize the potential of cloud computing in a secure environment. It is therefore obvious that FedRAMP will maintain resources to keep abreast of the technological and security enhancements in near real-time. As these cloud computing best practices evolve, FedRAMP security requirements, processes and templates will also undergo an evolution. The following sections define the FedRAMP change management process.

Factors for change

The following internal and external factors will drive the change to FedRAMP security requirements, processes and templates.

- ☐ Update to NIST special publications and FIPS publications: FedRAMP templates and requirements are based on the NIST special publications and FIPS publications. If the NIST SP 800-53 r3 is updated with new security controls and enhancements for low and moderate impact level, FedRAMP security controls will need to be updated. Also, if NIST publishes new guidance associated with cloud computing best practices, these will be considered for updates to FedRAMP security requirements and evaluation criteria/test procedures.
- ☐ Requirements from other Federal security initiatives: Government-wide security initiatives and mandates such as Trusted Internet Connections (TIC) and Identity, Credential and Access Management (ICAM) will drive updates to FedRAMP requirements for wider adoption of cloud computing systems and services across the Government. As the solutions for various cloud service models (IaaS, PaaS, SaaS), which is currently under active investigation, are adopted, they will be disseminated by FedRAMP. FedRAMP and the JAB will rely on both ISIMC and NIST to recommend changes to security controls over time. While these bodies will not have the authority to implement the changes, their expertise and reputation lend themselves to providing invaluable assistance to FedRAMP. It should be noted that security requirements can only be approved for change by the JAB.
- ☐ Agency-Specific requirements beyond the FedRAMP baseline: Federal Agencies leveraging FedRAMP authorizations for use within their own Agencies may add specific additional security controls, conduct additional assessments, or require implementation of other compensating controls. The leveraging agencies should notify FedRAMP of these additional requirements. FedRAMP JAB will meet regularly to discuss any required updates and possible inclusion of these additional security measures to FedRAMP security controls baseline and assessment procedures/evaluation criteria. If different leveraging agencies have added different requirements and additional security measures for the same cloud system, FedRAMP will maintain a list of these additions and may consider updating either the FedRAMP baseline for all cloud systems or just that specific cloud system. In both cases, FedRAMP will assess these additional controls/measures during the continuous monitoring phase.

- ☐ Industry best practices, development of standards or use of new tools/technology: FedRAMP requirements may be updated to adopt new standards as they are created for cloud computing interoperability, portability and security. As cloud computing market matures and as industry develops new tools and technologies for automated and near real-time monitoring of controls and automated mechanisms for exposing audit data to comply with regulatory requirements become available, FedRAMP processes will also be updated accordingly.
- ☐ Changes to cloud service provider offering: As new features and components are added to the cloud service provider offering, additional requirements and assessments might be necessary to ensure that robust security posture of the system is maintained.

Security Documents/Templates Change Control

All security document templates are to be considered "living documents". Over time, as requirements change, methodologies evolve, or new technologies and threats present themselves, these documents will undergo some degree of modification. FedRAMP is solely responsible for implementing these changes. It should be noted that FedRAMP security document templates are designed to assist the user with proper documentation related to their authorization package.

These also serve to provide a more uniform content collection method that aids the CSP and agencies with achieving authorization status for the cloud service offering. As changes are made, updated templates will be posted to the FedRAMP website with instructions related to use.

Requirements for Cloud Service Provider Change Control

Process

Once a requirement is approved, CSP have 30 days to develop and submit an implementation plan. CSP are responsible for implementing the plans. The implementation plan needs to define the actions that the CSP must perform in order to comply with the new requirement. In most cases the implementation of the new control will be implemented within the 30 day window. However, there may be instances where the implementation of the controls will require the CSP to add the control to the POAM sheet, with milestones, target date, and resource allocations documenting the future implementation due to the nature of the control itself.

Furthermore, it is understood that, depending on the particular infrastructure related to the security control, it might be necessary for the CSP to implement a compensating control. This control will accomplish the same goal as the new requirement. However, it accomplishes the goal in a different manner. All compensating controls must receive authorization from the JAB. When situations arise where the new requirement cannot be implemented on a system due to the legacy nature of the infrastructure, or in cases where the control itself will have a severely negative impact on the mission of the system, the CSP may request a waiver. Waivers, though rare, must be presented to the JAB for approval. Once the control change is implemented, FedRAMP is to be notified and the security control baseline will be adjusted and documented.

Sponsoring Agency CCP

Sponsoring federal agencies maintain their responsibility for establishing and maintaining their own internal change control process. Responsibilities related to the cloud computing service offering should be limited to the interconnection between the agency and the CSP, and the input to any change requests.

Appendix 3. 2010 Digital Universe Study

iView content – FINAL, Version: 4-26-2010

Title: A Digital Universe Decade – Are You Ready?

Tab 1: The Digital Universe Decade

"You Ain't Seen Nothing Yet." The title of that track from the 1974 Bachman-Turner Overdrive album Not Fragile aptly describes the state of today's Digital Universe. Between now and 2020, the amount of digital information created and replicated in the world will grow to an almost inconceivable 35 trillion gigabytes as all major forms of media – voice, TV, radio, print – complete the journey from analog to digital.

At the same time, the influx of consumer technologies into the workplace will create stresses and strains on the organizations that must manage, store, protect, and dispose of all this electronic content. So, if you have ever suffered from information overload or been bombarded with e-mails, texts, instant messages, documents, pictures, videos, and social network invitations, get ready, this is just the beginning.

Since 2007, on behalf of EMC Corporation, IDC has been sizing what it calls the Digital Universe, or the amount of digital information created and replicated in a year.

Here are just a few points to whet your appetite for the rest of the tabs in this IDC iView:

- ☐ Last year, despite the global recession, the Digital Universe set a record. It grew by 62% to nearly 800,000 petabytes. A petabyte is a million gigabytes. Picture a stack of DVDs reaching from the earth to the moon and back.
- ☐ This year, the Digital Universe will grow almost as fast to 1.2 million petabytes, or 1.2 zettabytes. (There's a word we haven't had to use until now.)
- ☐ This explosive growth means that by 2020, our Digital Universe will be 44 TIMES AS BIG as it was in 2009 (Figure 30). Our stack of DVDs would now reach halfway to Mars.

Figure 30. The Digital Universe 2009–2020

The Digital Universe 2009-2020

Growing by a Factor of 44

2009
.08 ZB*

*Zettabyte =
1 Trillion gigabytes

2020
35 ZB

Source: IDC Digital Universe Study, sponsored by EMC, May 2010

Here's another question for you. What comes after a quadrillion? That's right, a quintillion, an incomprehensible number, yet the one you need to describe the number of information containers – packets, files, images, records, signals – that the bits in the Digital Universe will be in by 2020. There will be 25 quintillion containers.

These containers, the files if you will, are the things that are actually managed, protected, and stored in the Digital Universe.

And, because of the growth of embedded systems in the smart grid, smart cities, logistic item tracking, and so on, the average file size in the universe is getting smaller. The number of things to be managed is growing twice as fast as the total number of gigabytes. Good luck, all you CIOs out there.

Think of the growth of the Digital Universe as a perpetual tsunami. As this universe grows by an order of magnitude, we will have to deal with information in new ways:

- ☐ How will we find the information we need when we need it? We will need new search and discovery tools. Most of the Digital Universe is unstructured data (for example, images and voice packets). We will need new ways to add structure to unstructured data, to look INSIDE the information containers and recognize content such as a face in a security video. In fact, the fastest-growing category in the Digital Universe is metadata, or data about data.
- ☐ How will we know what information we need to keep, and how will we keep it? Yes, we will need new technical solutions tied to storage, but we will surely also need new ways to manage our information. We'll need to classify it by importance, know when to delete it, and predict which information we will need in a hurry.

- ☐ How will we follow the growing number of government and industry rules about retaining records, tracking transactions, and ensuring information privacy? Compliance with regulations has become an entire industry – a $46 billion industry last year – but will it be enough?
- ☐ How will we protect the information we need to protect? If the amount of information in the Digital Universe is growing at 50% a year or so, the subset of information that needs to be secured is growing almost twice as fast. The amount of UNPROTECTED yet sensitive data is growing even faster.

As we contemplate the growth of the Digital Universe, these are some of the things we need to think about:

- ☐ New search tools
- ☐ Ways to add structure to unstructured data
- ☐ New storage and information management techniques
- ☐ More compliance tools
- ☐ Better security

There are plenty of others, including the role of cloud computing, the consumerization of the workplace, the growing share of the Digital Universe coming from China and India, and the growing diversity – in content type and container type – of the Digital Universe.

Here is another statistic to keep in mind before you start reviewing our other findings: Although the amount of information in the Digital Universe will grow by a factor of 44, and the number of containers or files will grow by a factor of 67 from 2009 to 2020, the number of IT professionals in the world will grow only by a factor of 1.4.

Big changes are coming.

Tab 2. Information in the Clouds

By 2020, a significant portion of the Digital Universe will be centrally hosted, managed, or stored in public or private repositories that today we call "cloud services." And even if a byte in the Digital Universe does not "live in the cloud" permanently, it will, in all likelihood, pass through the cloud at some point in its life.

There are almost as many definitions of cloud services as there are vendors trying to gain advantage by offering them. But in the IDC definition, they require availability over a network, consumption on-demand with pay-as-you-go billing, and some level of user control and system openness that separates cloud services from simple online delivery of content. It's software as a

service, not downloading software programs. It's watching on-demand internet TV, not merely downloading Netflix videos.

At the same time, cloud services can be offered as a shared common functionality (public cloud) or as a private version (private cloud), where an organization maintains complete control of all of the IT resources and how they are managed and secured. The latter can even be offered within the enterprise itself.

Although IDC has only sized the market for IT functionality – hardware, software, services – delivered over the public cloud at this point, a look at the makeup of the Digital Universe by content type – entertainment, financial transaction, medical information, and user-generated image content – gives one a feel for how big cloud services in other industries could become in the Digital Universe.

Using reasonable forecast assumptions such as those in the IDC forecast for IT cloud services, it is possible to conclude that as much as 15% of the information in the Digital Universe in 2020 could be part of a cloud service – created in the cloud, delivered to the cloud, stored and manipulated in the cloud, etc. Even more information could "pass through the cloud," that is, be transported using a cloud services e-mail system or shared community, be stored temporarily on disk drives in the cloud, be secured via a cloud service, etc. By 2020, more than a third of all the information in the Digital Universe will either live in or pass through the cloud (Figure 31).

Because so much of the content in the Digital Universe is in the form of images – from digital TV to user-generated images – nearly 50% of the Digital Universe cloud-based subset will be tied to entertainment.

Figure 31. The Digital Universe in the Clouds, 2020

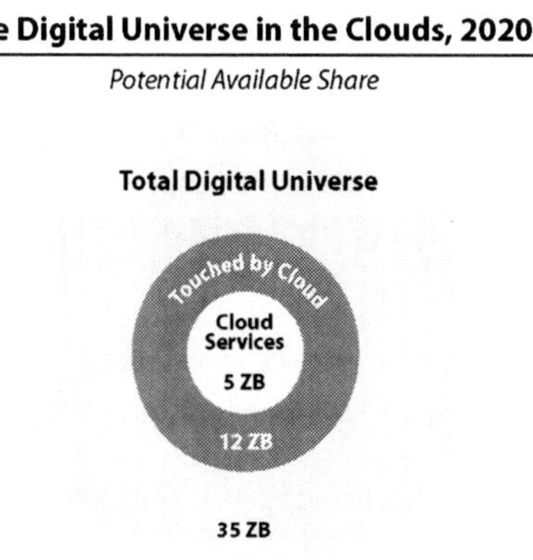

But every industry will have some form of cloud-based services to offer. And with these services come unique challenges and benefits:

- ☐ The promise of cloud services, besides lower upfront investment, is ease of management. But as users of IM systems in the early days and users of iTunes today realize, conversion from one service to another is not always seamless.
- ☐ Cloud services imply access to broadband services, which are not always available. Perhaps more challenging is the fact that cloud services may spur more need for bandwidth. Already, telco carriers are complaining that users of smart phones are using much more bandwidth than predicted as they rush to use more and more applications.
- ☐ Overall, security may actually be enhanced in the cloud – as cloud providers build security and transparency into the cloud infrastructure from the get-go. But the impact of a failure, should there be one, could be significant. This could be a problem in private clouds as well as public clouds. And with the loss of physical control over infrastructure, ensuring visibility in the cloud will be critical for demonstrating compliance.

Even with information stored or manipulated in the cloud, enterprises still have responsibility and liability over it. Managing this from afar might be a challenge.

But the benefits of cloud services will be substantial. Using IDC's current forecast for IT cloud services and assuming that the use of cloud services could lower the portion of the IT budget devoted to legacy system maintenance by a fraction of a percent, we estimate that the increase in IT dollars spent on innovation could drive more than $1 trillion in increased business revenues between now and the end of 2014.

Tab 3. Protected and Unprotected Data

Do you know where your social security number is?

Think about it. You probably have to enter it onto forms 10 times a year, maybe 50. From there it enters the Digital Universe, living in all sorts of databases, from those in your doctor's office and tax accountant's office, to bank records, company personnel records, mortgage records, and so on. It shuttles around from database to database at the speed of light. It gets backed up again and again.

In fact, those 10 entries (counting the entries into linked or associated files, accessed by people allowed to access the databases with your number in it, and the backups) could be propagated as many as a million times in a year!

How secure are those entries?

By 2020, almost 50% of the information in the Digital Universe will require a level of IT-based security beyond a baseline level of virus protection and physical protection. That's up from about 30% this year. And while the portion of that part of the Digital Universe that needs the highest level of security is small – in gigabytes and total files – that portion will grow by a factor of 100.

Not all data needs to be protected equally. A YouTube video of a cat doing tricks would seem to need less protection against hacking or corruption than a home-banking customer's account balances. But each YouTube video is associated with an IP address and end-user profile, and of course, that video might not be of a cat but of something not fit for public viewing. Even worse, that seemingly innocuous cat video may actually be an effective delivery mechanism for the new variants of malware being created by the criminal underground.

For the sake of understanding the degree of security in the Digital Universe, we have classified information that requires security into five successively higher security-level categories:

- ☐ Privacy only – such as an e-mail address on a YouTube upload
- ☐ Compliance-driven – such as e-mails that might be discoverable in litigation or be subject to retention rules
- ☐ Custodial data – account information, a breach of which could lead to or aid in identity theft
- ☐ Confidential data – information the originator wants to protect, such as trade secrets, customer lists, or confidential memos
- ☐ Lockdown data – information requiring the highest security, such as financial transactions, personnel files, medical records, or military intelligence

Obviously, information can switch categories over its life, or, in aggregation, gain more value over time and hence need higher security. The information that you visited a single website might be less sensitive than your entire web-browsing history, or even less sensitive than the information about how many times you visited the website.

By examining information by category and source, it's possible to estimate the amount of information in the Digital Universe that needs some level of security. That which doesn't is mostly transient data, especially digital TV signals and voice packets that aren't needed after a broadcast or call is over (Figure 32).

Figure 32. The need for information security

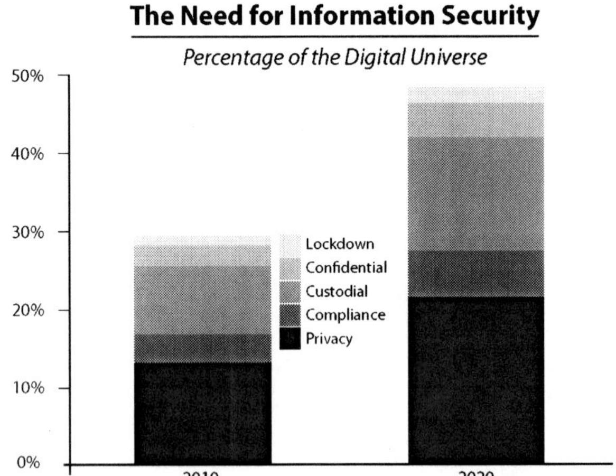

However, just because the information should have protection doesn't mean that it does. Using the same process of categorization, it is possible to estimate the amount of information that is not adequately protected.

And that amount is growing.

If you look at the information in the Digital Universe that needs to be protected by number of containers or "files" (rather than by number of bytes), the percentage needing protection is more than 90%. And the amount of unprotected data will grow by a factor of 90 between now and 2020 (Figure 33).

Figure 33. Unprotected data Needing Protection

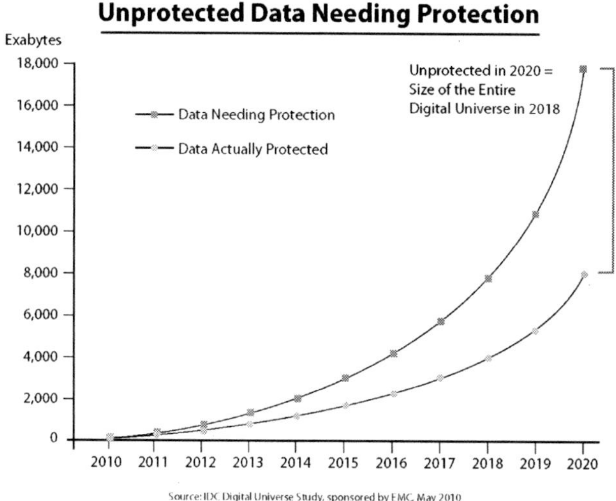

The issues for CIOs are pretty clear. They need access to tools and expertise to protect this burgeoning quantity of data needing protection in both the physical and, increasingly, the virtual worlds. But they also need the support of the business units in dealing with the policy and training issues involved – and getting that support has been a perennial problem. Without another regulatory driver such as Sarbanes-Oxley or a catastrophic breach, it is often difficult for CIOs to get the full attention of management regarding the non-technical aspects of information security.

In fact, the issue is even more complex. What a consumer or company wants protected (e-mails they wish to recall, buying or searching patterns, old Facebook photos) may change from day to day depending on circumstances or because of changes in the originator's own status. (For example, Sarah Palin's e-mail account in the state of Alaska got a lot more interesting to hackers the day she was announced as a Vice Presidential candidate, and it was hacked soon thereafter.)

Finally, they may not even know there is data about them in the Digital Universe. If they did know about it, they would want it to be protected or otherwise proscribed. In creating the model for the Digital Universe, IDC has discovered that the gigabytes a person may create through his or her own actions – taking photos, blogging, sending e-mails, getting cash from an ATM, downloading MP3s – is less than 10% the information ABOUT that person in the Digital Universe. The other 90% is composed of credit records, surveillance photos, analytics on behavior, web-use histories, and so on.

Probably EVERY byte in the Digital Universe could use some security and privacy protection. But we will never know because we can never know exactly what all those files and gigabytes actually contain.

Tab 4: The Future of Digital Information Storage

In the inaugural Digital Universe study, IDC forecasted that by 2007, for the first time, the amount of digital information created would exceed the amount of available storage. That inflection point has since become a gulf that continues to expand. Every year, the industry ships thousands of petabytes of new storage capacity, including hard disk drives, optical, tape, nonvolatile memory (flash), and volatile memory (DRAM). The total amount of storage capacity available is equal to all new shipments of storage plus all unused storage from previous years. While previously consumed storage could be overwritten, IDC assumes this capacity is reserved for what is already stored on the storage medium.

Hence, we have a growing gap between the amount of digital content being created and the amount of available digital storage. IDC estimates that in 2009, if people had wanted to store every gigabyte of digital content created, they would have had a shortfall of around 35%. This gap is expected to grow to more than 60% (that is, more than 60% of the petabytes created could not be stored) over the next several years (Figure 34).

Figure 34.The emerging gap

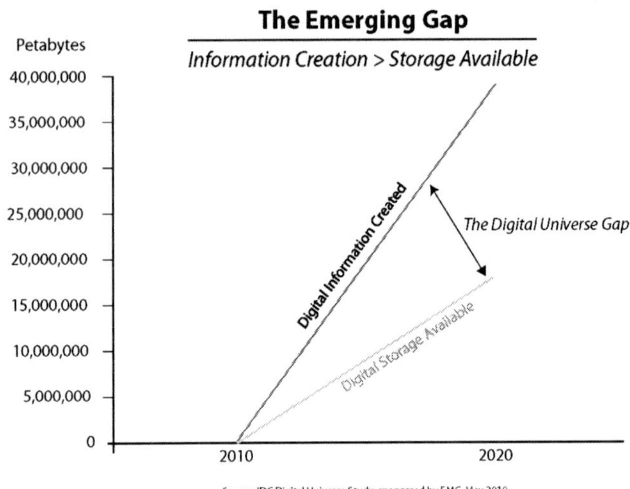

While much of the digital content we create is simply not that important (not much different from the paper magazines and newspapers that we throw away, or the telephone conversations,

receipts, bad pictures, etc., that we never save), the amount of data that does require permanent or longer-term preservation for a multitude of reasons is increasing exponentially.

But as we peer into the future, we see the greatest challenges are related not to how to store the information we want to keep, but rather to:

- ☐ Reducing the cost to store all of this content
- ☐ Reducing the risk (and even greater cost) of losing all of this content

Extracting all of the value out of the content that we save IDC data shows that nearly 75% of our digital world is a copy – in other words, only 25% is unique. Granted, various laws and regulations require multiple copies exist in order to ensure the availability of data over a long period of time. Multiple copies of data are also necessary to ensure proper performance by applications, or to protect data in the event of hardware failures. Nevertheless, the amount of data redundancy is excessive in many cases, and it represents a prime area for improvement and cost reduction.

Most of today's de-duplication happens on 2nd tier storage, but it doesn't have to. Tomorrow's opportunity is for de-duplication on primary storage (assuming no impact to performance), which would significantly reduce or eliminate post-process de-duplication. The cloud is an especially attractive place to eliminate redundancy, given its one-to-many model of content aggregation.

IDC's forecast portends a huge amount of consumer-attached storage, either directly to hosts or via a network. The external drive market will continue to experience considerable growth. Hence, there is a growing need to enable storage management by consumers. As computing races down the mobility path, one person may find himself with multiple computing devices (each potentially with its own operating system), and various amounts of local storage, all of them providing similar ways to access the cloud (via wireless cellular or wireless broadband).

Sharing the same content among these devices will increase in importance. The cloud must play a strategic role in becoming the central axis for all of this content – yet there is much work to do in order to make this happen. Until then, individuals will strive to keep passwords, content, and services mapped appropriately among their digital device portfolio.

Personalized services (e.g., GPS, proactive coupon pushing, e-commerce) will grow in importance as more data about one's personal habits and preferences is captured and mined by sophisticated applications. Granted, many people today may find this to be intrusive or a direct violation of their personal privacy. But as newer generations of teens and adults use their beloved digital devices, various types of data will be captured and leveraged to deliver services that not only will be embraced, but also will be considered one of those can't-get-along-without-it technologies. The personal data captured and used to deliver these services must be tied to an individual anonymously, and it must be managed in accordance with strict governance and compliance

procedures in order to ensure that privacy is not breached. This is, as they say, "easier said than done." Nevertheless, it must be done.

Personal privacy is paramount, and although there are behaviors one can employ to lessen one's digital footprint, no one can completely disappear off the ubiquitous digital grid. Video surveillance and purchasing transactions will be nearly impossible to avoid, and interaction in social networks will be either commonplace or a temptation that is difficult to overcome.

In the end, corporations and consumers will continue to create, copy, and store information –mostly on traditional storage technologies such as hard disk drives. Much of the data that is created will be disposable or a means to a final copy – hence the growing gap between content creation and available storage. However, the content that is stored will be deemed important and vital for evolving our businesses and business models, for extending our digital presence (that should result in more convenience and personalization), and perhaps most important, for protecting our digital heirlooms.

Tab 5: Consumers Without Borders

When you think of the Digital Universe, you may think of the financial databases of Wall Street, the acres of servers operating at giant internet service providers, or the storage devices supporting 100 million enterprises in the world.

But, in fact, most of the Digital Universe begins with an action by a consumer – an e-mail typed on a laptop, a digital photo taken at a wedding, a movie downloaded from Netflix.

In fact, more than 70% of the Digital Universe this year will be generated by users – individuals at home, at work, and on the go. That's 880 billion gigabytes.

At the same time, most of the gigabytes in the Digital Universe pass through the servers, network, or routers of an enterprise at some point. When they do, the enterprise is responsible at that moment for managing that content, protecting user privacy, watching over account information, and protecting copyright. It was the breach of personal e-mail accounts in China that drove Google off the mainland this year – an excellent example of enterprise liability for consumer-created data.

We classify user-generated content for which enterprises are responsible as "enterprise touch." About two thirds of all user-generated content falls into this category. Here's another way to think of it: While enterprise-generated content accounts for 20% of the Digital Universe, enterprises are liable for 80% (Figure 35).

Figure 35. User Creation = Enterprise worries

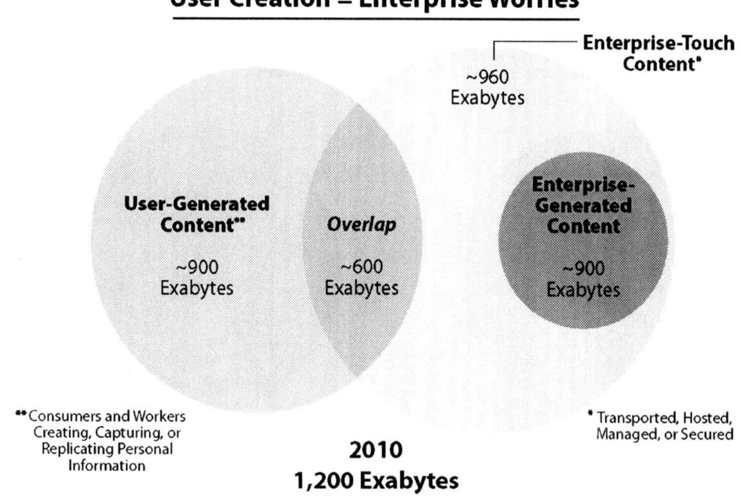

This enterprise liability will only get worse as social networking and Web 2.0 technologies continue to permeate the enterprise. Research by IDC shows that workers who use those technologies at home frequently also use them at work, often commingling their personal and business content. They use the same smart phones and laptops for work and personal activities. If they blog at home, they want to blog at work.

Many companies are struggling to keep up with this blurring of personal and enterprise boundaries. Only half of the companies surveyed in a separate study by IDC have any kind of corporate guidelines for employee use of social media at work. And nearly two thirds of the enterprise social media activity taking place is being driven by employee initiatives (Figure 36).

Figure 36. Social Networking Invades the Enterprise

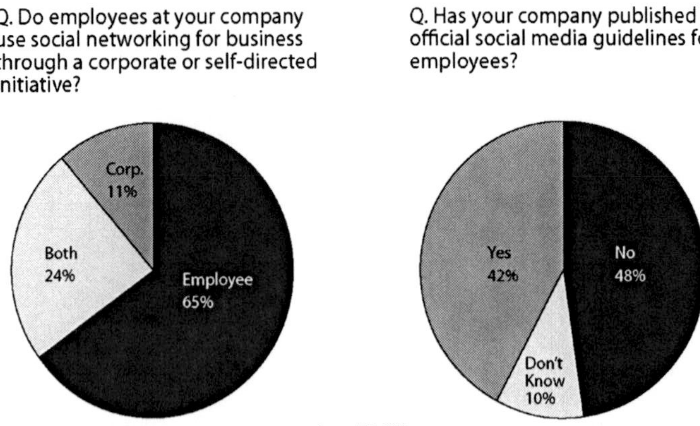

This blurring of boundaries is also blending the risks to information. As workers increasingly bring their personal devices and online habits to work, they are also bringing behind the firewall and into the corporate data center the viruses, Trojans, and other malware typically associated with consumer online fraud. This exposes the corporation to increased risk of information theft.

Employers experience several dilemmas in harnessing social media for the enterprise:

- ☐ Setting policy. Do you support all employee-created content and devices? Do you let any employee or department set up customer or supplier social media communities? If you set standards, how do you enforce them? Who sets the policy – individual departments, corporate headquarters, IT, business units?
- ☐ Assuming you see the need for enterprise social networking, how do you make it easy for individuals and departments to engage in Web 2.0 activities – do you offer 100% support or just point them in the right direction?
- ☐ How do you ensure employee or department social media information is backed up and archived properly, that it adheres to relevant standards and laws, and that it supports the enterprise brand?
- ☐ How do you ensure security, privacy, and intellectual property protection for these employee-driven activities?

These are all issues now being faced by enterprises. And the social media invasion of the enterprise has just begun. IDC estimates that by 2020, business transactions on the internet –business-to-business and business-to-consumer – will reach 450 billion a day.

The interactions taking place in the Digital Universe will not only add to its size and growth, but also to the increasing challenge of managing and securing it.

Tab 6. Bucks and Bytes

In 2009, the world spent nearly $4 trillion on hardware, software, services, networks, and IT staff to manage the Digital Universe. That spending is expected to grow modestly between now and 2020, which means the cost of managing each byte in the Digital Universe will drop steadily – an incentive to create even more information (Figure 37).

Figure 37. The decreasing Cost of Managing Information

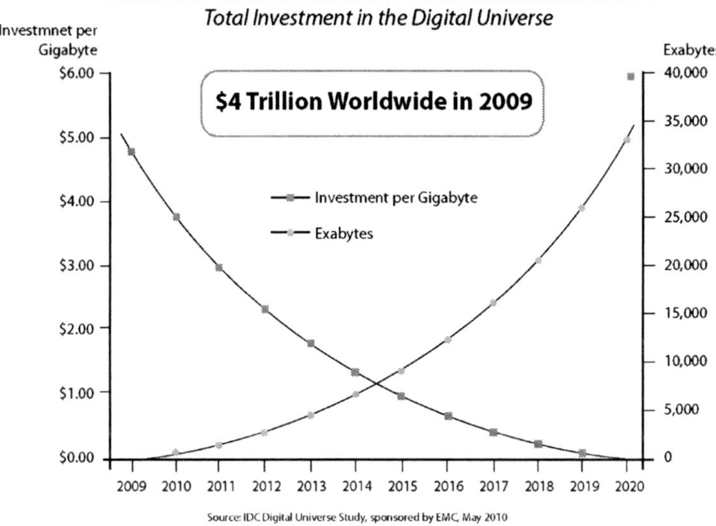

There are implications here. While the cost per byte drops, so does the investment in IT staff per byte. The investment in IT staff per information container, or file, drops even faster (Figure 38).

Figure 38. The decreasing Cost of Managing Information (IT)

This falling investment ratio, while spurring the growth of the Digital Universe, also means that the tools for managing it will have to change. For instance:

- ☐ The increased complexity of managing digital information will be an incentive to migrate to cloud services.
- ☐ Within data centers, expect continued pressure for data center automation, consolidation, and virtualization.
- ☐ For the management of end-user and customer business transactions, look for more end-user self-service.
- ☐ Expect bottlenecks in key specialties such as security (especially security), information management, advanced content management, and real-time processing.
- ☐ This manage-more-with-less situation will put ever-increasing stress on IT organizations. Those that manage this stress better than competitors will have an advantage.

Tab 7: Call to Action

Between 2009 and 2020, the information in the Digital Universe will grow by a factor of 44; the number of "files" in it to be managed will grow by a factor of 67, and storage capacity will grow by a factor of 30.

Yet the staffing and investment to manage the Digital Universe will grow by a factor of 1.4.

This should make it clear to CIOs and business executives that much of the next 10 years of their careers will be spent dealing with the challenge of the mismatch of these growth rates.

From reviewing the various tabs in this iView, we can narrow the CIO's issues down to:

- ☐ Developing tools for search and discovery of information as the Digital Universe expands, including finding ways to add structure to unstructured data though metadata, automatic content tagging, and pattern recognition.
- ☐ Deploying tools for new levels of information management and prioritized storage.
- ☐ Deploying tools and expertise for security and privacy protection for a growing portion of the Digital Universe in hybrid physical/virtual environments.
- ☐ Getting ready for some level of conversion to cloud-based services to start equipping IT staff with the new skills required for providing IT as a service and to obtain some economy of scale for ever-scarcer IT talent.
- ☐ Obtaining support from top management and from business units to implement the non-technical aspects of dealing with the Digital Universe, including setting policies on social media, training end-users on information security, and classifying information in an effort to set storage priorities.

As ever, the first order of business is to educate management about the issues and infuse a sense of urgency across the enterprise. As we saw during the recession of 2009, the Digital Universe is a force unto itself. It will grow whether enterprises are ready for it or not.

The second order of business is to plan more than one step out. For instance, if a major new storage management application takes two years to implement – from planning to infrastructure upgrade completion – the Digital Universe will be twice as big at the end of the project as it was at the beginning.

The third order of business is to consider issues of privacy, security, and protection early on and to embed solutions into developing infrastructures, particularly in the virtual space, where we have the opportunity to build-in security from the get-go.

The fourth order of business is to continue working on the relationship with the business units, where most of the funding for new projects originates and where most of the corporate desire to drive policy and training resides.

CIO Action Items

- ☐ Deploy new IT tools for information management and security.
- ☐ Develop a sense of urgency in top management and the business units.
- ☐ Develop a long-term plan.

- ☐ Increase the bond with the business units.
- ☐ Offload work to the cloud.

In many ways, consumers and employees have some of the same action items when it comes to dealing with their contributions to the Digital Universe. Chief among these is a sense of urgency when it comes to security. Also important is gaining an awareness of how much information about them sits in the Digital Universe and how fast that information is growing.

Appendix 4. Building Return on Investment from Cloud Computing

Computing Key Performance Indicators and Metrics

Cloud Computing introduces an expanded context for service-oriented business and IT.

Developing ROI models that show how Cloud Computing adoption can benefit both business and IT consumers and providers involves examining the key technology features and business operating model changes.

This section gives an overview of ROI models to support Cloud Computing assessments and business cases in two aspects:

- ☐ Key Performance Indicator ratios that target Cloud Computing adoption, comparing specific metrics of traditional IT with Cloud Computing solutions. These have been classified as cost, time, quality, and profitability indicators relating to Cloud Computing characteristics.
- ☐ Key Return on Investment savings models that demonstrate cost, time, quality, compliance, revenue, and profitability improvement by comparing traditional IT with Cloud Computing solutions.

The overview of Cloud Computing ROI models considers both indicators and ROI viewpoints.

Overview of Cloud Computing ROI models and KPIs.

Cloud Computing ROI Models and KPIs

Figure 39. Cloud Computing ROI models and KPIs

Cloud Computing ROI Models | **Cloud Computing KPIs**

Cloud Computing ROI Models				Cloud Computing KPIs			
	Speed of Reduction	Optimizing time to deliver/ execution	**Time**	Availability versus Recovery SLA	Workload-Predictable Costs	Workload-Variable Costs	Capex versus Opex Costs
Speed of Reduction	Optimizing cost of Capacity	Optimizing Ownership Use	**Cost**	Workload versus Utilization %	Workload type allocations	Instance to Asset ratio	Ecosystem-Optionality
	Optimizing cost to deliver/ execution	Green costs of Cloud	**Quality**	Experiential	SLA Response error rate	Intelligent automation	
		Optimizing Margin	**Margin**	Revenue Efficiencies	Market Disruption rate		

Figure 40. Cloud Computing ROI models –Cost Indicator ratios

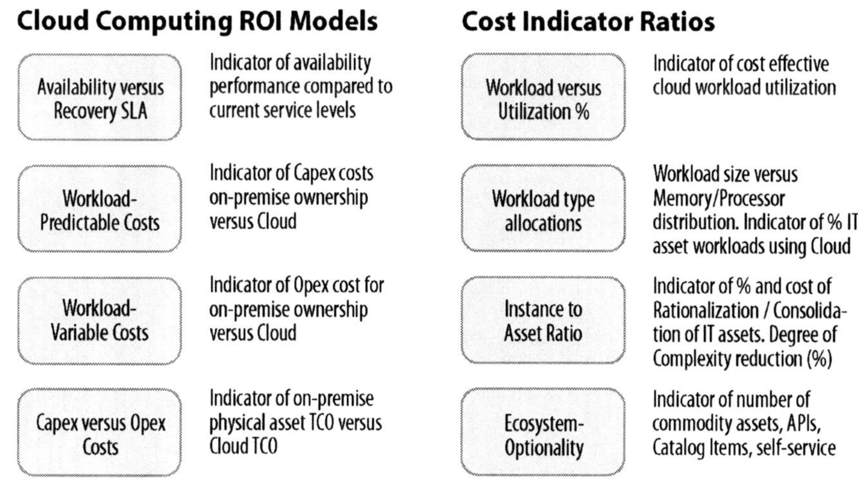

Availability versus recovery SLA:

- ☐ Indicator of availability performance compared to current service levels

Workload – predictable costs:

☐ Indicator of CAPEX cost on-premise ownership versus Cloud

Workload – variable costs:

☐ Indicator of OPEX cost for on-premise ownership versus Cloud; indicator of burst cost

CAPEX versus OPEX costs:

☐ Indicator of on-premise physical asset TCO versus Cloud TCO

Workload versus utilization %:

☐ Indicator of cost-effective Cloud workload utilization

Workload type allocations:

☐ Workload size versus memory/processor distribution; indicator of % IT asset workloads using Cloud

Instance to asset ratio:

☐ Indicator of % and cost of rationalization/consolidation of IT assets; degree of complexity reduction

Ecosystem – optionality:

☐ Indicator of number of commodity assets, APIs, catalog items, self service

Figure 41. Cloud Computing ROI Models – Time Indicator Ratios

Cloud Computing ROI Models - Time Indicator Ratios

Timeliness	The degree of service responsiveness An indicator of the type of service choice determination
Throughput	The latency of transaction The volume per unit of time throughout An indicator of workload efficiency
Periodicity	The frequency of demand and supply activity The amplitude of the demand and supply activity
Temporal	The even frequency to real-time action and outcome result

Timeliness:

☐ The degree of service responsiveness

☐ An indicator of the type of service choice determination

Throughput:

☐ The latency of transactions

☐ The volume per unit of time throughput

☐ An indicator of the workload efficiency

Periodicity:

☐ The frequency of demand and supply activity

☐ The amplitude of the demand and supply activity

Temporal:

☐ The event frequency to real-time action and outcome result

Cloud ROI Quality Indicator Ratios

☐ Quality indicator ratios.

Figure 42. Cloud ROI Quality Indicator Ratios

Cloud Computing ROI Models - Quality Indicator Ratios

Experiential	The quality of perceived User experience of the service Quality of User Interface design and interaction - ease of use
SLA Response error rate	Frequency of defective responses
Intelligent automation	The level of automated response (agent)

Cloud ROI Quality Indicator Ratios

Experiential:

☐ The quality of perceived user experience

☐ The quality of User Interface (UI) design and interaction – ease-of-use

SLA response error rate:

☐ Frequency of defective responses

Intelligent automation:

☐ The level of automation response (agent)

Cloud ROI Profitability Indicator Ratios

Profitability indicator ratios.

Figure 43. Cloud ROI Profitability Indicator Ratios

Cloud Computing ROI Models - Profitability Indicator Ratios

Revenue Efficiencies	Ability to generate margin increase per revenue Rate of annuity improvement
Market Disruption Rate	Rate of Revenue growth Rate of New Product market acquisition

Revenue efficiencies:

☐ Ability to generate margin increase/budget efficiency per margin

☐ Rate of annuity revenue

Market disruption rate:

☐ Rate of revenue growth

☐ Rate of new market acquisition

Cloud ROI Savings Models

The savings models.

Figure 44. Cloud Computing ROI Savings Models

Cloud Computing ROI - Savings Models

Time

Speed of Reduction

Rate of change of TCO reduction by Cloud adoption

Optimizing time to deliver / execution

Increase in Provisioning speed
Speed of multi-sourcing

Cost

Speed of Reduction

Rate of change of TCO reduction by Cloud adoption

Optimizing cost of Capacity

Aligning cost with usage. Capex to Opex
Utilization pay-as-you-go savings from Cloud adoption

Optimizing Ownership Use

Portfolio TCO License cost reduction from Cloud adoption
Open Source Adoption
SOA Resue Adoption

Quality

Green costs of Cloud

Green sustainability

Optimizing time to deliver / execution

Reduced supply chain costs
Flexibility / Choice

Profitability

Optimizing Margin

Increase in Revenue/ Profit margin from Cloud adoption

Speed of time reduction:

☐ Compression of time reduction by Cloud adoption

☐ Rate of change of TCO reduction by Cloud adoption

Optimizing time to deliver/execution:

☐ Increase in provisioning speed

☐ Speed of multi-sourcing

Speed of cost reduction:

☐ Compression of cost reduction by Cloud adoption

☐ Rate of change of TCO reduction by Cloud adoption

Optimizing cost of capacity:

☐ Aligning cost with usage, CAPEX to OPEX utilization pay-as-you-go savings from Cloud adoption

☐ Elastic scaling cost improvements

Optimizing ownership use:

☐ Portfolio TCO , license cost reduction from Cloud adoption

☐ Open Source adoption

☐ SOA re-use adoption

Green costs of Cloud:

☐ Green sustainability

Optimizing time to deliver/execution:

☐ Increase in provisioning speed

☐ Reduced supply-chain costs

☐ Speed of multi-sourcing

☐ Flexibility/choice

Optimizing margin:

☐ Increase in revenue/profit margin from Cloud adoption

Appendix 5. Achieving Operational Efficiency

As part of a broader IT transformation, the Federal Government needs to fundamentally shift its mindset from building custom systems to adopting light technologies and shared solutions. Too often, agencies build large standalone systems from scratch, segregated from other systems. These systems often duplicate others already within the Federal Government, wasting taxpayer dollars. The growth in data centers from 432 in 1998 to 2,094 in 2010 highlights this problem.

Leading private sector companies have taken great strides to improve their operating efficiencies. Cloud technologies and Infrastructure-as-a-Service enable IT services to efficiently share demand across infrastructure assets, reducing the overall reserve capacity across the enterprise. Additionally, leveraging shared services of "commodity" applications such as e-mail across functional organizations allows organizations to redirect management attention and resources towards value-added activities. The massive scale of the Federal Government allows for great potential to leverage these efficiencies.

Apply "Light Technology" and Shared Solutions

The shift to "light technologies," that is, cloud services, which can be deployed rapidly, and shared solutions will result in substantial cost savings, allowing agencies to optimize spending, and allowing agencies to reinvest in their most critical mission needs. For example, GSA recently entered into a contract to shift e-mail services to the cloud, resulting in a 50% cost reduction over five years – a savings of about $15 million. Agencies must focus on consolidating existing data centers, reducing the need for infrastructure growth by implementing a "Cloud First" policy for services, and increasing their use of available cloud and shared services.

1. Complete detailed implementation plans to consolidate at least 800 data centers by 2015

In February 2010, the Administration launched the Federal Data Center Consolidation Initiative (FDCCI) and issued guidance for Federal CIO Council agencies. The guidance called for agencies to inventory their data center assets, develop consolidation plans throughout FY 2010, and integrate those plans into agency FY 2012 budget submissions.

The FDCCI is aimed at assisting agencies in identifying their existing data center assets and formulating detailed consolidation plans that include a technical roadmap and clear consolidation targets. The FDCCI will cut down the number of data centers across the government and assist agencies in applying best practices from the public and private sector, with goals to:

- ☐ Promote the use of Green IT by reducing the overall energy and real estate footprint of government data centers
- ☐ Reduce the cost of data center hardware, software, and operations
- ☐ Increase the overall IT security posture of the government, and
- ☐ Shift IT investments to more efficient computing platforms and technologies.

After an 8 month peer review process, we now know that the government is operating and maintaining almost 2,100 data centers. Through the FDCCI, a minimum of 800 data centers will be closed by 2015.

To meet this reduction target, OMB and agency CIOs will take the following steps:

1.1 Identify agency data center program managers to lead consolidation efforts

Large IT projects often fail to meet goals because of distributed accountability for success. Large, complex, and critical infrastructure programs, such as data center consolidation, require a single person to lead the coordinated effort.

Within the next six months, each agency will designate a senior, dedicated data center consolidation program manager with project management experience and technical competence

in IT infrastructure. Because data center consolidation requires interactions with many stakeholder groups, the data center program manager must also have strong communication skills. The data center program manager at each agency will be responsible for developing a plan with interim, verifiable milestones to reach the agency's data center reduction target and monitor progress toward those goals.

1.2 Launch a Data Center Consolidation Task Force to ensure successful execution

Within the next three months, the Federal CIO Council will launch a government-wide Data Center Consolidation Task Force comprised of the data center program managers, facilities managers, and sustainability officers. The Data Center Consolidation Task Force will be responsible for working together to share progress toward individual agency goals and the overall Federal target of a minimum of 800 data center closures by 2015.The Data Center Consolidation Task Force will meet monthly to review progress of each consolidation project and ensure government-wide alignment between agency efforts where appropriate. The Task Force will serve as a "community of practice" for agency CIOs and data center program managers to share best practices from this effort and enhance consolidation effectiveness.

1.3 Launch a publicly available dashboard to track data center consolidation progress

OMB will launch a publicly available dashboard to serve as a window into progress of the data center consolidation program. The dashboard will ensure transparency and accountability, and keep the overall program in plain view of the public.

2. Create a government-wide marketplace for data center availability

Within the next 18 months, OMB and GSA will create a government-wide marketplace that better utilizes spare capacity within operational data centers.

This online marketplace will match agencies with extra capacity to agencies with increasing demand, thereby improving the utilization of existing facilities. The marketplace will help agencies with available capacity promote their available data center space. Once agencies have a clear sense of the existing capacity landscape, they can make more informed consolidation decisions.

3. Shift to a "Cloud First" policy

In the private sector, a web-based multimedia production company used the cloud to allow anyone with access to an internet connection the ability to create their own fully customized, professional-quality, TV-like videos. Consumers can then share the resulting videos with friends and family across the world. The cloud allowed for a rapid response when demand jumped from 25,000 users to more than 250,000 users in three days, eventually reaching a peak rate of 20,000

new customers every hour. Because of the cloud, the company was able to scale from 50 to 4,000 virtual machines in three days to support increased demand on a real-time basis.

In contrast, the Federal Government's Car Allowance and Rebate System (CARS, more commonly known as "Cash-For-Clunkers") failed when faced with peak loads. To process the anticipated 250,000 transactions, the National Highway Traffic Safety Administration (NHTSA) deployed a customized commercial application hosted in a traditional data center environment on June 19, 2009.When dealer registrations began on July 24, 2009, demand far exceeded initial projections, and within three days, the system was overwhelmed, leading to numerous unplanned outages and service disruptions. Ultimately, approximately 690,000 CARS transactions were processed. However, lacking the ability to scale rapidly, system stability was not achieved until August 28, 2009, over a month after registrations started coming in.

The Federal Government must be better prepared in the future. Beginning immediately, the Federal Government will shift to a "Cloud First" policy.

The three-part strategy on cloud technology will revolve around using commercial cloud technologies where feasible, launching private government clouds, and utilizing regional clouds with state and local governments where appropriate.

Cloud computing brings a wide range of benefits:

- ☐ Economical: Cloud computing is a pay-as-you-go approach to IT, in which a low initial investment is required to begin, and additional investment is needed only as system use increases.
- ☐ Flexible: IT departments that anticipate fluctuations in user demand no longer need to scramble for additional hardware and software. With cloud computing, they can add or subtract capacity quickly and easily.
- ☐ Fast: Cloud computing eliminates long procurement and certification processes, while providing a near-limitless selection of services.

When evaluating options for new IT deployments, OMB will require that agencies default to cloud-based solutions whenever a secure, reliable, cost-effective cloud option exists. To facilitate this shift, we will be standing up secure government-wide cloud computing platforms.

3.1 Publish cloud strategy

Within the next six months, the Federal CIO will publish a strategy to accelerate the safe and secure adoption of cloud computing across the government.

The National Institute of Standards and Technology (NIST) will facilitate and lead the development of standards for security, interoperability, and portability. NIST is working with other

agencies, industry, academia, standards development organizations, and others to use existing standards as appropriate and develop cloud computing standards where gaps exist. While cloud computing services are currently being used, experts cite security, interoperability, and portability as major barriers to further adoption. The expectation is that standards will shorten the adoption cycle, enabling cost savings and an increased ability to quickly create and deploy enterprise applications.

3.2 Jump-start the migration to cloud technologies

Each agency CIO will be required to identify three "must move" services and create a project plan for migrating each of them to cloud solutions and retiring the associated legacy systems. Of the three, at least one of the services must fully migrate to a cloud solution within 12 months and the remaining two within 18 months.

Each migration plan will include major milestones, execution risks, adoption targets, and required resources, as well as a retirement plan for legacy services once cloud services are online. These new cloud implementations should be compatible with the secure, certified platforms currently provided in the private sector. Migrating these services will build capabilities and momentum in the Federal Government, encourage industry to more rapidly develop appropriate cloud solutions for government, and reduce operating costs.

4. Stand-up contract vehicles for secure IaaS solutions

Federal, state, and local governments will soon have access to cloud-based Infrastructure-as-a-Service (IaaS) offerings. GSA's IaaS contract award allows 12 vendors to provide government entities with cloud storage, virtual machines, and web hosting services to support a continued expansion of governments' IT capabilities into cloud computing environments.

Within the next six months, after completing security certification, GSA will make a common set of contract vehicles for cloud-based Infrastructure-as-a-Service solutions available government-wide.

A government-wide risk and authorization program for cloud computing will allow agencies to rely on the authorization completed by another agency or to use an existing authorization, so that only additional, agency-specific requirements need to be separately certified. The aim is to drive to a set of common services across the government supported by a community, rather than an agency-specific risk model. This will allow the Federal Government to "approve once and use often."

5. Stand-up contract vehicles for commodity services

The Software-as-a-Service (SaaS) E-mail Working Group, formed in June 2010, has begun to identify and develop the set of baseline functional and technical requirements for government-wide cloud e-mail solutions and is working towards developing business case templates for agencies who are considering transitioning to SaaS e-mail.

Within 12 months, GSA will utilize these requirements to stand up government-wide contract vehicles for cloud-based e-mail solutions. GSA will also begin a similar process specifically designed for other back-end, cloud-based solutions.

6. Develop a strategy for shared services

Within the next 12 months, the Federal CIO will develop a strategy for shared services. That strategy will build on earlier Federal Government successes in shared services and include benchmarks on current usage and uptake rates, as well as service level agreements (SLAs), customer satisfaction levels, costs, and overall economic effectiveness.

Managing partners of shared services will assess the current state of shared services and each release a roadmap to improve quality and uptake. Ultimately, the managing partners will be responsible for executing these roadmaps and will be held accountable for improvements on SLAs and reductions in cost. These efforts will enable the current shared services to be accessible government-wide at higher quality levels.

Action item owner and deadlines

	Action Item	Owner	Within 6 months	6–12 months	12–18 months
1	Complete detailed implementation plans to consolidate 800 data centers by 2015	OMB, Agencies	X		
2	Create a government-wide marketplace for data center availability	OMB, GSA			X
3	Shift to a "Cloud First" policy	OMB, Agencies	X		
4	Stand-up contract vehicles for secure IaaS solutions	GSA	X		
5	Stand-up contract vehicles for "commodity" services	GSA		X	
6	Develop a strategy for shared services	Federal CIO		X	

Effectively managing large-scale IT programs

IT has transformed how the private sector operates and has revolutionized the way in which it serves its customers. The Federal Government has largely missed out on these transformations, due in part to its poor management of large technology investments.

To address these execution problems, it launched the IT Dashboard in June 2009, allowing the American people to monitor IT investments across the Federal Government and shining a light onto government operations. While this unprecedented transparency was an important first step, it was not enough to simply shine a light on problems and hope that solutions would follow.

Building on the foundation of the IT Dashboard, we launched TechStat Accountability Sessions ("TechStats") in January 2010.A TechStat is a face-to-face, evidence-based review of an IT program with OMB and agency leadership. TechStat sessions enable the government to turnaround, halt, or terminate IT investments that do not produce dividends for the American people.

As a result of more than 50 TechStat reviews, OMB now has a sharper picture of the persistent problems facing Federal IT. One of the most consistent problems lies in project scope and timeline. In TechStat sessions, OMB found that many current IT projects are scheduled to produce the first deliverables years after work begins, in some cases up to six years later. In six years, technology will change, project sponsors will change, and, most importantly, program needs will change. Programs designed to deliver initial functionality after several years of planning are inevitably doomed.

Modular development delivers functionality in shorter timeframes and has long been considered best practice in the private sector and in some areas of government; in fact, both Raines Rules and the Federal Acquisition Regulation (FAR) advise agencies to plan programs in this way. Successful organizations using modular development base releases on requirements they define at a high level and then refine through an iterative process, with extensive engagement and feedback from stakeholders. To maintain the discipline of on-time and on-budget, organizations push out additional functionality and new requirements for major changes into future releases and prioritize critical needs and end-user functionality.

Evidence shows that modular development leads to increased success and reduced risk. However, because this is a new way of thinking about IT programs for some groups within government, it requires additional training, templates, and tools. Many existing government processes – from planning to budgeting to procurement – naturally favor larger, more comprehensive projects. As such, far too many Federal IT programs have multi-year timeframes well beyond the now accepted 18- to 24-month best practice. The activities outlined in this plan

attempt to address the structural barriers to implementing modular development consistently across government.

Moving forward, Federal IT programs must be structured to deploy working business functionality in release cycles no longer than 12 months, and, ideally, less than six months, with initial deployment to end users no later than 18 months after the program begins.

Program managers need to define each phase of the IT development lifecycle and rigorously manage scope. These timelines should encompass the entire process – from concept through requirement analysis, development, test, and delivery. Today, a number of agencies have implemented these modular practices successfully. The Department of Veterans Affairs now requires that large IT programs deliver working functionality every six months.

The following practices will help achieve the promises of modular development:

- ☐ Ensuring each module aligns with overall program and business objectives and has clear quantitative and qualitative outcome measures for success
- ☐ Awarding contracts that incorporate clear business objectives and performance outcomes, a vision for future state architecture, and parameters for iterative design and development
- ☐ Delivering new working functionality to users at least every 12 months, with no more than 3 months dedicated to creating detailed system specifications
- ☐ Regularly capturing and incorporating user feedback through an iterative process that assesses user satisfaction with each release, continuously refining design to ensure alignment with business needs
- ☐ Preventing scope creep by defining high-level requirements upfront, locking down the current release, and pushing additional non-critical functionality to future releases
- ☐ Moving resources from one release phase to the next as soon as they complete their work (e.g., the requirements team builds requirements for the next release, while developers build current release)

Strengthen Program Management

Effectively managing modular IT programs requires a corps of program and project management professionals with extensive experience and robust training. Strong program management professionals are essential to effectively steward IT programs from beginning to end, align disparate stakeholders, manage the tension between on-time delivery and additional functionality, and escalate issues for rapid resolution before they become roadblocks. The size and criticality of large Federal Government IT programs are considerable. The people managing these programs must represent the best of the best.

Challenges with program management are pervasive across the Federal Government due to a general shortage of qualified personnel. However, pockets of excellence exist in the government. For example, the Social Security Administration (SSA) has a developed a multi-tier career track for program managers that requires both training and experience for advancement. Program managers advance by gaining experience on small projects before moving to larger, more complex programs. SSA feels so strongly about the critical role of program managers that it will not begin a new program unless the right manager is in place and dedicated to lead it.

High-performing IT organizations have a well-developed program management talent strategy. The Office of Personnel Management (OPM), working with the Chief Human Capital Officers Council, will need to take steps to significantly enhance the supply of IT program management talent in the Federal Government. Steps include creating a career path to attract and reward top performers, establishing integrated, multi-disciplinary program teams with key skills before beginning major IT programs, requiring program managers to share best practices at the close of each program, launching a technology fellows program, and encouraging mobility of program managers across the government.

7. Design a formal IT program management career path

In the next six months, OPM, with input from agencies and OMB, will create a specialized career path for IT program managers (PMs).This will likely require creating a separate Occupational Series specific to IT program management within the current IT family with career advancement paths that are more competitive with the private sector. The path should require expertise and experience for advancement. It will also require the development of a competency model for IT program management consistent with the IT project manager model.

Finding, recruiting, and hiring top IT program management talent is challenging. In the next six months, OPM will work with OMB to provide agencies with direct hiring authority for IT PMs as necessary.

Further, agencies will identify specific IT program management competency gaps in the next Human Capital Management Report and develop specific plans to close the IT PM gap. To ensure that agencies are executing these plans, senior agency executives will review their progress and provide an interim report to OMB, 12 months after the next Human Capital Management Report is published.

OPM will work with the Department of the Treasury and the Department of Agriculture (USDA) to pilot the IT program management career track.

8. Scale IT program management career path government-wide

After piloting IT program management career paths at Treasury and USDA, OPM will work to expand the IT program management career paths more broadly across the Federal Government.

9. Require integrated program teams

A primary challenge impacting the successful delivery of IT programs is the need to manage a broad set of stakeholder communities, including agency leaders, business process owners, IT, acquisition, financial management, and legal. The typically siloed nature of government stakeholder communities is ill-suited for the multi-disciplinary and rapidly evolving needs of major IT program management processes.

High-performing private sector firms quickly bring together small multi-disciplinary, integrated program teams (IPTs) consisting of the following functions: business process owners who have a clear vision of the problem they are solving, IT professionals who understand the full range of technical solutions, acquisition professionals who plan and procure needed labor and materials, and finance staff to secure required funding. In addition, other functions such as HR and legal are included on the program team as needed. At the hub of these IPTs is a strong and effective program manager who stewards the process from beginning to end.

Examples of high-functioning IPTs exist in pockets of the Federal Government in which a complete IPT is required for major programs prior to beginning the investment review process. However, the practice is still only unevenly applied. The healthcare.gov initiative at the Department of Health and Human Services provides a good example of what a fully integrated multi-disciplinary team can do in the Federal Government. The healthcare.gov team successfully launched a citizen-facing website within 90 days of program initiation to rave reviews.

Over the next six months, OMB will issue guidance requiring an IPT, led by a dedicated, full-time program manager and supported by an IT acquisition specialist, be in place for all major IT programs before OMB will approve program budgets.

9.1 Dedicate resources throughout the program lifecycle and co-locate when possible

For each large IT program, critical members of the IPT will serve as full-time resources dedicated to the program. This must include a 100% dedicated IT program manager, but other roles will vary by program. Key members of the IPT will also be co-located during the most critical junctures of the program. This is especially important during the requirements-writing phase, when business, IT and acquisition must define and modify requirements in short iterative cycles, and when "translation issues" have historically caused problems.

The core of the IPT, including all IT program leadership roles, will be in place throughout the program lifecycle, from the initial concept development phase through the delivery of the last increment under the contract. For major IT investments, agency leadership will approve the composition of the integrated program team and the dedicated program manager.

9.2 Agencies will hold integrated program team members accountable for both individual functional goals and overall program success

A pervasive issue in government programs is that individual stakeholders focus primarily on performance metrics within their functions, and not on the holistic outcomes of the program. For example, IT or program staff may push to award work to a particular vendor, or to add "bells and whistles" that fail to take into account time pressures and budgetary constraints. Similarly, contracting staff may focus so much on competition requirements and small-business participation goals that they fail to look for solutions that meet these important requirements while also satisfying program needs. We need to replace these "stove piped" efforts, which too often push in inconsistent directions, with an approach that brings together the stakeholders and integrates their efforts.

Agency executives will work with their senior procurement executives (SPEs), CIOs, and program leaders to take action and drive towards a more balanced set of individual and program success metrics based on the following two recommendations:

- ☐ First, agencies should set up individual performance goals that cover individual and program objectives. Performance goals for acquisition, IT, and business personnel need to include a combination of individual and program objectives.
- ☐ Second, agencies must also ensure that the individual and program metrics balance speed, quality, effectiveness, and compliance with Federal Acquisition Regulations. Supervisors must utilize a balanced set of performance metrics to evaluate individual performance. Individuals who provide exemplary contributions to the team will be recognized for their success (e.g., acquisition recognition through the Federal Acquisition Institute Awards & Recognition Program for individuals who effectively meet program needs without sacrificing compliance).

10. Launch a best practices collaboration platform

Within six months, the Federal CIO Council will develop a collaboration portal to exchange best practices, case studies, and allow for real-time problem solving. To institutionalize this best practice sharing, agency PMs will submit post-implementation reviews of their major program deliveries to the portal. These reviews will populate a searchable database of synthesized and codified program management best practices that all PMs can access.

11. Launch technology fellows program

Within 12 months, the office of the Federal CIO will create a technology fellows program and the accompanying recruiting infrastructure. By partnering directly with universities with well-recognized technology programs, the Federal Government will tap into the emerging talent pool and begin to build a sustainable pipeline of talent. The technology fellows programs should specifically target competency gaps that are identified in the Human Capital Management Reports submitted by agencies.

The program will aim to cut bureaucratic barriers to entering public service and provide access to unique career opportunities in government agencies. At the same time, these roles will provide new fellows with relevant training in large IT program management.

12. Enable IT program manager mobility across government and industry

The Federal CIO Council, OMB, and OPM, over the next 12 – 18 months, will be responsible for developing a process that will support and encourage movement of program managers across government and industry. Rotational opportunities allow the Federal Government to leverage its size to share knowledge and expertise across agencies.IT program managers with experience on specific types of programs or with specific types of systems should have opportunities to apply this experience on similar programs across government. Similarly, program managers should be given opportunities to learn from leading private companies. The Federal CIO Council, OMB, and OPM will work to design opportunities for industry rotation to allow Federal program managers to remain up-to-date with the latest skills while managing conflict of interest issues.

To support PM mobility, the Federal CIO Council will build a repository of information on all Federal Government IT PMs, including relevant background, specific expertise, implementation experience, and performance as part of its best practices collaboration platform.

Action item owner and deadlines

	Action Item	Owner	Within 6 months	6-12 months	12-18 months
7	Design a formal IT program management career path	OPM, OMB,	X		
8	Scale IT program management career path	OPM, Agencies			X
9	Require Integrated Program Teams	OMB	X		
10	Launch a best practices collaboration platform	Federal CIO Council	X		
11	Launch technology fellows program	Federal CIO		X	

	Action Item	Owner	Within 6 months	6-12 months	12-18 months
12	Enable IT program manager mobility across government and industry	OMB, CIO Council, OPM			X

Align the Acquisition Process with the Technology Cycle

The acquisition process can require program managers to specify the government's requirements up front, which can be years in advance of program initiation. Given the pace of technology change, the lag between when the government defines its requirements and when the contractor begins to deliver is enough time for the technology to fundamentally change, which means that the program may be outdated on the day it starts.

The procurement reforms enacted in the 1990s provided tools to speed up the acquisition process, but the government has failed to take full advantage of those tools, so we continue to see programs delayed longer than the life of the technology. In particular, the use of multiple-award indefinite-delivery, indefinite-quantity (ID/IQ) contracts, called for in the 1994 Federal Acquisition Streamlining Act (FASA), was intended to allow quicker issuance of task orders, to be competed through streamlined "fair opportunity" mini-competitions among the multiple contract holders. The creation of government-wide acquisition contracts (GWACs) for purchasing IT goods and services was also intended to provide a limited number of specialized vehicles open to the entire government that could quickly respond to individual agency needs.

While the innovations in FASA have produced benefits, too often those tools are not used or not used effectively.IT acquisition, particularly for large projects, continues to move intolerably slowly.We need to make real change happen, by developing a cadre of specialized acquisition professionals and by educating the entire team managing IT projects about the tools available to streamline the acquisition process.

In addition, requirements are often developed without adequate input from industry, and without enough communication between an agency's IT staff and the program employees who will actually be using the hardware and software. Moreover, agencies often believe that they need to develop a cost estimate that is low in order to have the project approved. As a result, requirements are too often unrealistic (as to performance, schedule, and cost estimates), or the requirements that the IT professionals develop may not provide what the program staff expect – or both. Speeding up the acquisition timeline and awarding more successful contracts for IT requires a multifaceted set of solutions including increased communication with industry, high functioning, "cross-trained" program teams, and appropriate project scoping.

13. Design and develop a cadre of specialized IT acquisition professionals

Effective IT acquisition requires a combination of thorough knowledge of the Federal acquisition system, including the tools available, a deep understanding of the dynamic commercial IT marketplace, and the unique challenges inherent to successfully delivering large IT programs in a modular time-boxed manner. Agency CIOs and SPEs advised that acquisition professionals who were specialized in IT were more effective. This specialization is also consistent with private sector best practice. To bring these increased capabilities online, we will be creating standardized training and development opportunities to develop a cadre of acquisition professionals with the specialized knowledge and experience required to expedite complex IT acquisitions across the Federal Government.

Over the next six months, the Office of Federal Procurement Policy (OFPP) and the Federal CIO, with input from agencies, will design a specialized IT acquisition cadre. In doing so, they will need to answer the following questions:

- ☐ What is the process for acquisition professionals to become specialized in IT?
- ☐ How do professionals progress within the community (i.e., transition from entry-level through to senior contributor)?
- ☐ How do you ensure that community members can focus on participating in IT acquisition?
- ☐ What training, experience, and certification are needed?
- ☐ What will be the impact on the remaining acquisition workforce and non-IT acquisitions if some of the staff are dedicated to IT acquisition?

A number of agencies have already developed IT acquisition specialists who can serve as a means to expedite IT programs. Useful lessons can be learned from drawing on the experience of the GWACs and the staff that support them at GSA, NASA, and the National Institutes of Health (NIH).

In the case of smaller agencies, where IT-only acquisition groups may be impractical, leveraging GWACs or using specialized cadres at larger agencies through Economy Act transactions may be the best solution (e.g., the Department of Veterans Affairs' Technology Acquisition Center and Treasury Department's BPD Acquisition Resource Center).In addition, both the GWACs and these other agencies can potentially provide cross-functional support through experienced IT program management and technical staff. Access to these resources will, of course, not be limited to smaller agencies, as they can often provide an efficient alternative to in-house IT acquisition even for larger agencies. Particularly within the current budgetary constraints, agencies may have only a limited capability to hire new staff as candidates for the IT cadre, so drawing on other agencies' resources may be vital to success.

13.1 Strengthen IT acquisition skills and capabilities

Within six months, OFPP, with input from agencies, will develop guidance on requirements for IT acquisition specialists. In addition, OFPP will develop guidance on curriculum standards to cross-train program managers and IT acquisition professionals.

In particular, the guidance will focus on increasing cross-functional knowledge of the IT marketplace, IT program management, and IT acquisition. OFPP will build upon its current Federal Acquisition Certification in Contracting (FAC-C) to develop a path for IT expertise. OFPP will leverage existing curriculum that may exist within agencies that already host specialized IT acquisition professionals. Skills development will include:

- Classroom training: OFPP will leverage and strengthen, where necessary, existing classes at the Federal Acquisition Institute (FAI) and the Defense Acquisition University (DAU), and engage these and other training providers to develop additional offerings as necessary.
- On-the-job experience: As is true with acquisition in general, the skills needed to successfully handle large IT acquisitions call for a blend of classroom training and on-the-job experience. For example, contracting professionals with hands-on IT experience are better equipped to help IT and program staff translate business and technical requirements into a statement of work that can help ensure a smooth procurement.
- Mentorship: Building a strong culture of mentorship enables IT acquisition professionals to more quickly learn "the art of the possible" to deliver effective IT acquisition solutions. OFPP can encourage this by building on FAI's ongoing efforts to foster mentorship and networking opportunities, within and between agencies.

As an immediate action to implement these recommendations, OFPP will consider these initiatives as part of its review of the Federal Acquisition Certifications for Program/Project Managers (P/PMs), Contracting Officer's Technical Representatives (COTR), and contracting professionals in the next six months.

14. Identify IT acquisition best practices and adopt government-wide

OFPP will lead an effort over the next six months to study the experience of those agencies that have already created specialized IT acquisition teams, in order to develop a model to scale more broadly. Among the key questions to be considered will be the length of time individuals need to spend devoted solely to IT acquisition in order to add value to IT program teams, the kind of training and experiences that are most valuable, appropriate organizational structures, and successful acquisition strategies and practices.

Drawing on that experience, OFPP should work closely with senior agency leadership at the Department of Homeland Security (DHS) and Department of Energy (DOE) as they rollout their

IT acquisition cadres in the next year. The next step, over the following 18 months, is to scale the specialized IT acquisition cadre government-wide.

15. Issue contracting guidance and templates to support modular development

Over the next year, OFPP will work with the acquisition and IT communities to develop guidance on contracting for modular development. As part of this effort, OFPP will hold an open meeting with industry leaders to solicit ideas/feedback on contracting for modular development. OFPP will develop templates and samples, and will create communities of practice to facilitate adoption of modular contracting practices.

This guidance will address a variety of factors that IT program managers as well as contracting officers will need to consider as they plan for modular development efforts, such as whether to award to a single vendor or multiple vendors; how to ensure that there is appropriate competition at various stages in the process; how broad or specific the statements of work should be; when to use fixed-price contracts or rely on other pricing arrangements; and how to promote opportunities for small business. As noted above, the Federal Acquisition Streamlining Act of 1994 provides a variety of flexibilities for acquiring commercial items and for streamlining competition that will be reflected in the guidance to ensure IT program managers and others are aware of existing authorities that can further support modular IT development.

When evaluating acquisition strategies, agencies will need to prioritize those solutions that promote short deadlines for deliverables (generally less than three months), allow for responsiveness to rapidly evolving program and technical requirements, and facilitate a streamlined award process. One innovative example is at the Department of Transportation (DOT), which has recently formed IT "Agility Platforms" with contract vehicles in place that simplify how business owners can quickly access technology.

16. Reduce barriers to entry for small innovative technology companies

Small businesses in the technology space drive enormous innovation throughout the economy. However, the Federal Government does not fully tap into the new ideas created by small businesses. Unlike larger, more established firms, new entrants have little at stake in current technological systems ranging from software standards, to operating system and file standards, to business processes. While large firms drive many incremental improvements to the status quo, smaller firms are more likely to produce the most disruptive and creative innovations. In addition, with closer ties to cutting edge, ground-breaking research, smaller firms often have the best answers for the Federal Government.

However, small businesses too rarely approach the Federal Government as a customer because of the real and perceived barriers to contracting. The sales process is perceived as lengthy and

complex, and, therefore, not seen as worthwhile unless done at scale. Without existing knowledge or access to specialized lawyers and lobbyists, small firms default to more traditional channels. And given their limited size, small businesses often find it difficult to bid on the large chunks of government work that require a substantial workforce across many functional capabilities. Ultimately, the government contracting process is easier to navigate by large, existing players, who in turn dominate the volume of contracts and therefore create a track record making them "less risky" and more likely to win future contracts.

To address the barriers that small businesses face generally (both in IT and more broadly), in April 2010, the President established an interagency task force to make recommendations for improving the participation of small companies in Federal contracts. The task force made 13 recommendations in its August 2010 report, which are currently in various stages of implementation. Of the 13 recommendations, six were also included, in whole or in part, in provisions of the recently-enacted Small Business Jobs Act of 2010.

As part of this effort, and to enable small IT companies to work with the Federal Government, SBA, GSA, and OFPP will take concrete steps over the next 18 months to develop clearer and more comprehensive small business contracting policies.

Action item owner and deadlines

	Action Item	Owner	Within 6 months	6–12 months	12–18 months
13	Design and develop cadre of specialized IT acquisition professionals	OMB, Agencies	X		
14	Identify IT acquisition best practices and adopt government-wide	OFPP	X		
15	Issue contracting guidance and templates to support modular development	OFPP		X	
16	Reduce barriers to entry for small innovative technology companies	SBA, GSA, OFPP			X

Align the Budget Process with the Technology Cycle

The rapid pace of technological change does not match well with the Federal government's budget formulation and execution processes. In addition, modular development means that lessons learned from an early cycle in an IT program will likely inform the detailed plans for the next cycle. As such, agencies need more flexibility to manage IT programs responsibly. To compensate for this misalignment between the realities of IT program management and the need for detailed budgets several years in advance, several agencies have worked with Congress to achieve greater IT budget flexibility through multi-year and/or agency-wide portfolio appropriations.

To deploy IT successfully, agencies need the ability to make final decisions on technology solutions at the point of execution, not years in advance. Agencies need the flexibility to move funding between investments or projects within their portfolio to respond to changes in needs and available solutions. But at the same time, Congress has a legitimate and important need for oversight; and given the history of project failures and wasted investments, it is understandable that Congress requires compliance with a rigid system for managing IT investments.

The Department of Veterans Affairs (VA) presents an interesting model. Greater budget flexibility has allowed the VA CIO to freeze projects that are off track and either restructure them for success or cancel them.VA established an accountability system so projects that are missing milestones are flagged early. Greater budget flexibility paired with real-time visibility is leading to success at VA – and minimizing the risk of "big bang" failures.

17. Work with Congress to develop IT budget models that align with modular development

Working with Congress to design ways to better align funding to the technology cycle will reduce waste and improve the timeliness and effectiveness of provided solutions. Creating and leveraging flexible IT budget models requires work by OMB, Congress, and agency leadership.

17.1 Analyze working capital funds and transfer authorities to identify current IT budget flexibilities

Over the next six months, OMB will work with Congress to analyze existing working capital funds (WCFs) and other vehicles for pooling funds and extending availability of funding. Working capital funds (WCFs) are agency revolving funds for managing common administrative services that add budgeting flexibility within the agency. In addition to WCFs, franchise funds and other accounts can potentially provide added IT funding flexibility. These accounts add flexibility by pooling bureau-level funds to serve agency-wide purposes.

This analysis will address limits on the amount of funding that could flow through such accounts under current law across all appropriations and agencies, any limits on the types of activities that may be funded, and any other limitations on the use of transfer authorities to feed such accounts from contributing accounts. This analysis would also include a comprehensive review of the legislative language for accounts receiving funds such as WCF accounts, General Provisions, or other legislative limits on transfer authorities, and the legal limits on use of general transfer authorities such as the Economy Act or the E-Government Act of 2002.The analysis will also identify examples of the use of the existing funding flexibility vehicles for IT projects and develop best practices guidance on applicability and implementation across the government, as well as identify where skill gaps exist in developing costing models and managing funds.

17.2 Identify programs for which to pilot flexible budget models

Within six months, agency CIOs and CFOs will identify programs at several agencies for which added budget flexibility could save money and improve outcomes. OMB and agencies will work with Congress to develop proposed budget models to complement the modular development approach. In addition, OMB and agencies will evaluate mechanisms for increased transparency for these programs.

18. Develop supporting materials and guidance for flexible IT budget models

In order to support agencies and appropriations staff in leveraging budget flexibility, the Federal CFO Council, in collaboration with the Federal CIO Council, will develop a set of best practices and materials that explain the need for these types of funding, and prescribe a path to achieving more flexible models.

As a first step, the Federal CIO Council will create a segmentation of common IT program types and the associated funding requirements. The Federal CFO Council will then work with the Federal CIO Council to create detailed "playbooks" mapping each IT program type to specific budget vehicles based on examples of past investments and IT needs (e.g., multi-year funding for programs with several discrete deliverables).The playbooks will also explain in detail how the recommended budget flexibility improves delivery of the corresponding IT program results. Agencies will utilize these templates and training to clearly outline their financial needs to successfully deliver IT programs.

Program leaders and CIOs with increased budget flexibility will face higher expectations around successful delivery from agency leaders and Congress. Achieving greater flexibility in funding also requires greater transparency into spending effectiveness. Agencies will need to engage in more frequent dialogues with appropriations staff and to clearly demonstrate the performance of IT investments in achieving mission goals.

The Federal CFO and CIO Councils will create a set of guidelines for increasing transparency in the utilization of IT funds. Agencies will follow these guidelines and institute additional review processes for multi-year funds and portfolio funding to prevent mismanagement of increased funding flexibility (e.g., masking program delays or overruns).

19. Work with Congress to scale flexible IT budget models more broadly

Within 12 months, OMB will engage several agencies to work with Congress to launch flexible IT budget models where appropriate. As pilot agencies demonstrate success with flexible IT budget models on selected programs, OMB will continue to work with Congress to scale flexible budget models across major IT programs government-wide.

20. Work with Congress to consolidate commodity IT spending under Agency CIO

Agencies, departments, bureaus, and, at times, even programs currently design, build, and operate independent systems for "commodity" IT services (e.g., e-mail, data centers, content management systems, web infrastructure).Their functionality and the infrastructure that supports them are often duplicative and sub-scale. These independent systems currently draw resources away from IT programs that deliver value to the American taxpayer. With few exceptions, the minor differences between agency-specific systems and their associated operational processes do not drive value for the agencies.

Consolidating these systems and their associated infrastructure (e.g., data centers) will be difficult and complex if the current funding models are maintained. Within the next six months, OMB will work with Congress to develop a workable funding model for "commodity" IT services. These funding models will be applicable to both inter-agency IT services and intra-agency IT services. On an annual basis, the agency CIOs and the Federal CIO Council will identify "commodity" services to be included in this funding model as they are migrated towards shared services.

A benefit of consolidated commodity IT spending is the ability to move more rapidly to adopt strategic sourcing solutions. Once agencies with common business needs can effectively coordinate or consolidate the procurement of IT-related goods and services and demand is aggregated within agencies, it will be easier for the government to more effectively negotiate for volume discounts and improved service levels.

Action item owner and deadlines

	Action Item	Owner	Within 6 months	6–12 months	12–18 months
17	Work with Congress to create IT budget models that align with modular development	OMB, Agencies			
18	Develop supporting materials and guidance for flexible IT budget models	OMB, CFO Council, CIO Council		X	
19	Work with Congress to scale flexible IT budget models more broadly	OMB, Agencies			
20	Work with Congress to consolidate Commodity IT spending under agency CIO	OMB, Agencies	X		

Streamline Governance and Improve Accountability

To strengthen IT governance, we need to improve line-of-sight between project teams and senior executives, increase the precision of ongoing measurement of IT program health, and boost

the quality and timing of interventions to keep projects on track. These improvements will both boost the efficiency of project oversight and better manage programs in distress.

Our strategy for strengthening IT governance centers on driving agency adoption of the "TechStat" model currently used at the Federal level. TechStat Accountability Sessions are face-to-face, evidence-based reviews of agency IT programs with OMB and agency leadership. Using data from the Federal IT Dashboard, investments are carefully analyzed with a focus on problem-solving that leads to concrete action to improve performance.

TechStats have led to accelerated deliverables, budget reductions, and project terminations. Results include:

- ☐ $3 billion reduction in lifecycle costs
- ☐ Average acceleration of deliverables from over 24 months to 8 months

Our goal is to scale this capability across the Federal Government, increasing the number of programs that can be reviewed and hastening the speed at which interventions occur. Through this strategy, we aim to enable agencies to grow their own performance management standards and focus OMB direct involvement on a limited number of highest-priority cases.

21. Reform and strengthen Investment Review Boards

Investment Review Boards (IRBs) were created to control and evaluate the results of all major IT investments. In practice, these review boards have frequently failed to adequately manage the IT program portfolio by establishing successful projects or taking corrective action. Today, typical IRB meeting agendas currently set aside two hours to review the entire IT portfolio, far too little time to adequately review dozens of technical projects. These IRBs will be restructured according to the "TechStat" model.

21.1. Revamp IT Budget Submissions

OMB Exhibits 53 and 300 have come to support stand-alone processes to request and justify funding rather than serving as management tools for monitoring program health. In many cases, these documents are prepared in large part by third-party contractors and there is minimal involvement by agency executives and program managers.

These exhibits will be revamped to better align them to agency budgeting and management processes, make them more relevant and useful, and ensure they promote the use of modular development principles. The improved exhibits will also alleviate reporting burden, increase data accuracy, and serve as the authoritative management tool.

By May of 2011, OMB will reconstruct the 300s and 53s around distinct data elements that drive value for agencies and provide the information necessary for meaningful oversight. The timing of these elements will be separated into distinct streams to clarify objectives, give agencies adequate time to assemble strong responses, and improve data quality. These streams will include:

- ☐ Budget justification for new major Development, Modernization and Enhancement (DME) investments, significant re-engineering of existing DME investments, and annual re-justification of DME investments.
- ☐ Health monitoring of existing DME investments and Operations and Maintenance (O&M)
- ☐ Portfolio governance to ensure the IT portfolio and individual projects are consistent with the agency mission and Federal policy objectives

Importantly, OMB and agencies must evaluate the way in which IT programs are reviewed so that budget approval for large IT programs is tied to key implementation steps rather than seemingly upfront, wholesale approval of massive programs. OMB will evaluate ways to ensure agencies can demonstrate strong performance in earlier modules in order to receive approval for funding of subsequent modules.

21.2 Rollout "TechStat" model department-wide

By March 2011, OMB will work with agency CIOs and other agency leaders to stand up the "TechStat" model at the departmental level. Steps include:

- ☐ OMB will assist agencies in designing tools and enforcing their use, to provide the transparency required for the "TechStat" model to be effective
- ☐ OMB analysts will provide in-person training to agency CIOs in "TechStat" methodology including accountability guidelines, engagement cadence, evaluation processes, and reporting processes.
- ☐ Agency leaders will lead, sponsor, and manage the process within their departments

22. Redefine role of Agency CIOs and Federal CIO Council

Currently, agency CIOs and the Federal CIO Council spend a majority of their mindshare on policymaking and maintaining IT infrastructure. As we move forward with the IT reforms, CIO focus must shift towards portfolio management. This shift will be encouraged by activities such as the restructuring of the Investment Review Boards. Similarly, agencies will be increasingly freed from low-value activities (e.g., building redundant infrastructure) as they adopt technologies such as cloud computing.

- ☐ Agency CIOs will be responsible for managing the portfolio of large IT projects within their agencies. This portfolio management role will include continuously identifying unmet needs to be addressed by new projects, terminating or turning around poorly performing projects, and retiring IT investments which no longer meet the needs of the organization. Steps will include:

☐ As described above, agency CIOs will take on responsibility for the "TechStat" governance process within their agencies as of March 2011.

☐ Agencies will turnaround or terminate at least one-third of poorly performing projects in their portfolio within the next 18 months. The Federal CIO Council will play a similar portfolio management role, but at a cross-agency level. Within six months, the Federal CIO Council will periodically review the highest priority "TechStat" findings assembled by the agency CIOs. These reviews will enable CIOs to share best practices and common sources of failure to improve success rates over time.

23. Rollout "TechStat" model at bureau-level

Once cemented at the agency level, the "TechStat" model will be deployed at the bureau-level to ensure the effective management of large programs. Within 18 months, agency CIOs, in collaboration with other agency leaders, will be responsible for deploying the tools and training necessary to ensure rollout has been completed.

Action item owner and deadlines

	Action Item	Owner	Within 6 months	6–12 months	12–18 months
21	Reform and strengthen Investment Review Boards	OMB, Agencies	X		
22	Redefine role of agency CIOs and Federal CIO Council	Federal CIO, Agency CIOs	X		
23	Rollout "TechStat" model at bureau-level	Agency CIOs			X

Increase Engagement with Industry

The Federal Government does not consistently leverage the most effective and efficient available technologies. Federal IT contracts have been difficult to manage because they were not well-defined or well-written. These contractual challenges produce waste, delay program delivery, and erode the value of IT investments.

In many cases, agencies have been hindered by inadequate communication with industry, which is often driven by myths about what level of vendor engagement is permitted. The result has been barriers between industry and government buyers, whose efforts are often frustrated by a lack of awareness of the most efficient and effective technologies available in the private sector. These barriers negatively affect the full breadth of the acquisition process including needs identification, requirements definition, strategy formulation, the proposal process, and contract execution. Educating the community on the myths of vendor engagement will increase constructive and responsible engagement with the private sector IT community and improve the quality and cost effectiveness of the IT services provided.

24. Launch "myth-busters" education campaign

Commonly-held misunderstandings about how industry and government can engage with one another during the acquisition process place an artificial barrier between Federal agencies and their industry partners. These myths reduce the government's access to necessary market information as government officials, both program managers and contracting officers, are often unsure how to responsibly engage with their industry counterparts. They may have inaccurate information about the rules, may be overly cautious in their interactions, or may be unaware of communication strategies that can help the government define its requirements and establish sound acquisition strategies. The fact is that the statutory and regulatory framework for communications between industry and government allows significantly greater engagement than current practice. The government therefore needs to raise awareness of these flexibilities to its workforce.

OFPP will identify the major myths that most significantly hinder requirements definition and the development of effective acquisition planning and execution. In January 2011, OFPP will issue a memorandum identifying these myths and the related facts and strategies to improve constructive engagement. This effort will be supported through discussions and other outreach efforts with key stakeholders in early 2011 including, but not limited to:

- ☐ Professional associations and other industry representatives
- ☐ Federal stakeholders including program managers, contracting professionals, agency attorneys, and ethics officials

Throughout 2011, the Federal Acquisition Institute (FAI) and OFPP will conduct a "myth-buster" awareness campaign to eliminate artificial private sector engagement barriers. Steps will include at least the following:

- ☐ Launch an online community of practice within the next six months using technologies such as video channels to provide a Q&A forum, celebrate successes, and share "myths" and potential "myth-busters"
- ☐ Conduct FAI webinar for the acquisition workforce hosted by OFPP by late January 2011
- ☐ Create mandatory, continuous learning program through the FAI website
- ☐ Present at conferences such as the GSA Expo, the National Contract Management Association (NCMA) World Conference, and NCMA Government Contract Management Conference throughout 2011

25. Launch interactive platform for pre-RFP agency-industry collaboration

The government benefits when there is broad engagement with industry before beginning an IT project. Recently, the government used an online wiki tool to rapidly and effectively explore solutions for a planned Federal IT investment. Tens of thousands of visitors participated from

all 50 states and workers at Fortune 500 companies interacted with the owners of a 10-person business to discuss the best solutions for the government. The dialogue allowed participants to tag and vote on the best ideas, providing the agency with a list of top priorities and key themes that made the feedback both more comprehensive and more actionable than what could have been obtained through traditional methods. Technological opportunities were discussed, weighed, and judged by the community that was not immediately obvious at the onset of the effort.

Inexpensive, efficient solutions such as these should be made available to all agencies to effectively tap the understanding of industry partners, especially in the period prior to issuing a Request for Proposal (RFP).Within the next six months, GSA will launch a government-wide, online, interactive platform for this purpose. Action item owner and deadlines

Action item owner and deadlines

	Action Item	Owner	Within 6 months	6–12 months	12–18 months
24	Launch "myth-busters" education campaign	OFPP	X		
25	Launch an interactive platform for pre-RFP agency-industry collaboration	GSA	X		

Summary

From delivering benefits to our veterans to advancing biomedical discovery, Federal Government IT investments are designed to serve the American people. By focusing on execution, oversight, and transparency, this plan will deliver tangible results to stakeholders across the Federal Government and the American taxpayers.

Individually and together, the 25 actions detailed will move the government towards the future – more nimble, more cost effective, and more citizen-focused. These IT reforms require collaboration with Congress; engagement with industry; and commitment and energy from government leadership and IT, acquisition, and financial management professionals. They require relentless focus on near-term execution, recognition of past lessons, and a long-term vision for the future. But these efforts are worth the hard work. By shifting focus away from policy and towards execution and oversight, these IT reforms will succeed in delivering results for the American people.

The future picture for Federal Government IT is exciting.IT enables better service delivery, enhanced collaboration with citizens, and dramatically lower costs. We must get rid of the waste and inefficiencies in our systems. Outdated technologies and information systems undermine our efficiency and threaten our security.

Federal IT projects will no longer last multiple years without delivering meaningful functionality. Poorly performing projects will be identified early and put under a spotlight for turnaround – those that continue to flounder will be terminated. No longer will large IT contracts be negotiated by individuals without IT expertise. No longer will one agency build expensive new data centers when other agencies have excess capacity. And no longer will rigid budgeting constraints prevent executives from making smart decisions with taxpayer dollars; flexible models will allow agency leaders to shift funds where and when they are needed, ensuring that results matter more than plans.

A government powered by modern information technology is a faster, smarter, and more efficient government. While IT projects throughout the government will always have risks, there are no excuses for spectacular failures. And while not all projects can be perfect, major errors must and will be caught early and addressed appropriately. Projects should never be so far behind schedule that the primary activity of program managers shifts to waging a constant public relations battle to ensure continued funding. Instead, with streamlined governance and experienced program managers, issues can be caught early and course corrections can be made without wasting time and money.

The Federal Government will be able to provision services like nimble start-up companies, harness available cloud solutions instead of building systems from scratch, and leverage smarter technologies that require lower capital outlays. Citizens will be able to interact with government for services via simpler, more intuitive interfaces.IT will open government, providing deep visibility into all operations. With this 25 point plan, the Federal Government will turn the corner on implementing the most critical reforms, ensuring that large IT programs perform as expected and can be delivered on time and on budget in order to deliver for the American people.

Appendix 6. NIST Cloud Computing Business Use Case Template

1. Use Case Identification

1.1. Use Case Name

☐ State a concise, results-oriented name for the use case.

1.2. Agency

☐ Record the agency sponsoring this use case.

1.3. Model Matrix

☐ Identify which intersections of the service/deployment matrix the use case addresses.

2. Change History

Version Number	Author Name	Change Date	Section Changed	Description of Change

☐ The change table tracks the change history for the use case.

2.1. Version Number

☐ The current version of the working document.

2.2. Author Name

☐ Supply the name of the person who performed the most recent update to the use case.

2.3. Change Date

☐ Enter the date on which the use case was created or modified.

2.4. Section Changed

☐ Enter the section of the report that was changed.

2.5. Description of Change

☐ Provide a short description of the change or changes made to the document.

3. Background

☐ An abstract describing the purpose of the business use case.

4. Definitions

☐ Definitions specific to terms used in the business use case that require explanation.

Appendix 7. Cloud Computing Business Use Case - VDI

1. Use Case Identification

Use Case Identification

Use Case Name	Virtualized Desktop Infrastructure (VDI)				
Service / Deployment Matrix:			Service Model		
			Cloud Software as a Service (SaaS)	Cloud Platform as a Service (Paas)	Cloud Infrastructure as a Service (Iaas)
	Deployment Model	Private Cloud			X
		Community Cloud			X
		Public Cloud			
		Hybrid Cloud			X
Created By:		Last Updated By:			
Date Created:		Date last Updated:			

2. Primary Actors

- ☐ **Cloud-provider:** An organization providing network services and charging cloud-subscribers.
- ☐ **Cloud-subscriber:** A person or organization that has been authenticated to a cloud and maintains a business relationship with a cloud.
- ☐ **Cloud-subscriber-user:** A user of a cloud-subscriber organization who will be consuming the cloud service provided by the cloud-provider as an end user. For example, an organization's employee who is using a virtualized desktop on a thin client service the organization subscribes to would be a cloud-subscriber's user.

3. Business Goal

The main goals of a Virtual Desktop Infrastructure implementation are to decrease support costs through increased configuration consistency and simplified patch management and to increase accessibility of work environments by making virtualized environments available via web browser on any device. Security can be improved by requiring that applications be increased through the virtual desktop, which would necessitate the use of network storage.

3.1. Service Model

The service model considered in this use case is Infrastructure as a Service (IaaS) . In IaaS, the cloud-subscriber is able to deploy and run any number and type of virtualized desktops. The cloud-subscriber does not manage the underlying infrastructure, which is handled by the cloud-

provider. For this specific use case, the cloud-subscriber-user is able to access and authenticate to the cloud and access and use his virtual desktop from any device, including mobile devices.

3.2. Deployment Model

Three different deployment models are applicable to this use case. From the perspective of the cloud-provider, this is a Private cloud implementation. The primary motivation is to provide a secure cloud environment for deployment of virtual desktops. The first deployments will use an on-premise solution, and the cloud-provider and the cloud-subscriber will be the same.

As the agency gains experience, the cloud will become a Community cloud, shared by several organizations within the sponsoring department. This sharing will be driven by common concerns, such as security requirements, policy requirements, and compliance considerations. Once this occurs, cloud-subscribers will include those additional agencies, and the cloud-provider will be a single unit within the department.

The final iteration will see a Hybrid cloud deployment; a combination of community and private clouds, depending on security requirements. Agencies within the department that have greater security needs can obtain private clouds, while those with lower requirements would be on a Community cloud.

4. Necessary Conditions

4.1. Security

A means of authentication that uniquely identifies a cloud-subscriber-user and applies appropriate roles and privileges is necessary. A means of migrating security policies from existing systems is required.

- ☐ Data must be accessible by only those cloud-subscriber-users who should be allowed to.
- ☐ Security should be manageable at the same or better granularity as is currently possible.

Compliance with security policies should be enforced through the desktop to the degree possible and appropriate. For example, thin and zero client deployments can prevent the storage of any data on external devices.

4.2. Interoperability

The ability to deploy a VDI solution utilizing several different vendors (e.g. Citrix XenDesktop vs VMWare View) is highly desirable. The ability to use heterogeneous solutions is limited, both vertically (the various product layers between the cloud-provider and the cloud-subscriber-user) and horizontally (competing VDI solutions do not run on one another).

4.3. Portability

The cloud-provider needs to be able to migrate the cloud-subscriber-user's existing applications and data to the virtual desktop for the solution to be useful. In this particular use case, the cloud-provider also needs the ability to migrate from product to product in the VDI solution space (for example, moving from VMWare to Microsoft and vice versa).

Legacy application support is less problematic, as the various applications used by desktop users can be given their own virtual machine to run in. A problem that remains is that if the legacy application depends on other applications, then all must share the same virtual machine.

4.4. Other

There are several pieces to delivering a VDI solution, but most vendors are compatible with only a small number of other vendors. These pieces include:

- ☐ Virtualization platform.
- ☐ Communications protocol.
- ☐ Virtual management platform
- ☐ Session broker
- ☐ Client device

5. Priorities and Risks

The priorities of the VDI implementation primarily center on decreased operating costs and increased security. Improved security permits mobile computing in an increased number of cases, most generally because data are stored on the network rather than on the device. Increased end-user compliance with security policies (such as restrictions on the use removable media or installation of third-party software) is another significant benefit of VDI in environments with restricted data.

As stated previously, decreased operating costs are expected in the long run. This is due to improvements in patch management (and associated security improvements), standardized desktops, improved license management, and potentially less expensive end-user devices.

The same improvements that decrease operating costs can also improve end user support through faster troubleshooting of problems when they arise. Since there are few variations in the standard desktop, there are fewer one-off problems that need resolution. Additionally, when issues arise that impact multiple users, identified fixes can be propagated rapidly and efficiently throughout the organization.

Risks can also make the experience painful for the cloud-subscriber-user. In general, performance and service expectations for all devices and all locations will be driven by the service

and performance encountered on standard clients. Availability of the client will be compared to having a laptop. Decreased opportunity to modify the desktop or install desired applications will negatively affect perception.

The VDI infrastructure will need to be fully redundant to allow end-users to continue work without interruption, especially if any of them use thin or zero clients. Network bottlenecks need to be eliminated, and the networks themselves must be both high bandwidth and low latency. If used on a large scale, the entire infrastructure needs to be able to handle a surge of logins and desktop requests in a short period of time as people arrive at work. If thin clients or zero clients are used, then the performance and service must reach the level provided by a regular client.

Even as data backup capabilities increase due to increased use of network storage, backups will become more critical as no work would be conductible in the event of an outage. One identified risk is that there will be unconstrained growth on the os disk, especially in installations where each cloud-subscriber-user is mapped to their own virtual machine.

6. Essential Characteristics

NIST has identified five essential characteristics of cloud solutions.

- ☐ **On-demand self-service:** the cloud-subscriber can automatically provision additional cloud-subscriber-users without requiring human interaction by the cloud-provider.
- ☐ **Broad network access**: accessed through standard mechanisms on heterogeneous thin and thick clients. Both high bandwidth and low latency are expected.
- ☐ **Resource pooling:** multi-tenant model.
- ☐ **Rapid elasticity:** can be quickly scaled out and in, and additional cloud-subscriber-user licenses can be purchased in any quantity at any time.
- ☐ **Measured service:** both the cloud-provider and the cloud-subscriber must be able to automatically monitor and control resource use to provide transparency.

7. Normal Flow

The cloud-subscriber purchases and provisions licenses from the cloud-provider.

The cloud-subscriber monitors usage and metrics automatically provided by the cloud-provider.

The cloud-subscriber-user logs in at the business location using a client and is able to commence working after an interval indistinguishable from a normal boot of a PC.

The cloud-subscriber-user logs in from a remote location using a personally owned client and is able to commence working after an interval indistinguishable from a normal boot of a PC.

The cloud-subscriber-user logs in from a mobile device and is able to access and work on important files using applications from any type of network. Speed and latency are dependent on the host network, but security expectations remain the same.

The cloud-subscriber-user logs in and uses a device such as an iPad to access and work on important projects seamlessly.

8. Frequency of Use

From the perspective of the cloud-provider, all scenarios will be encountered daily. From the perspective of all but the smallest Federal cloud-subscribers, all scenarios will be encountered daily. The cloud-subscriber-user is most likely to log in daily from the work site, several times per week from their home computer, and occasionally using a mobile device while travelling.

9. Special Requirements

An appropriate broadband connection with low latency and high bandwidth is generally required for the best experience for the cloud-subscriber-user. When using a mobile client, lower capability network connections should be acceptable.

10. Notes and Issues

Supported for different types of clients must be determined; each may have its place. Type 1 versus Type 2 versus thin client versus zero client.

Will persistent connections be mandatory, implying that all activity occurs on servers in the data center, or will synchronization with the data center be adequate? The latter suggests that offline and disconnected clients are feasible, with additional complexity from the security perspective. However, disconnected clients are a solution when network bandwidth is limited.

Disconnected mode issues:

- ☐ Validating connecting user
- ☐ Virtual hd delivery
- ☐ Hypervisor delivery
- ☐ Client management
- ☐ Active session management
- ☐ Virtual hd synchronization
- ☐ Client/server vs local app support
- ☐ Secure endpoint and virtual desktop

Currently, the availability of tools used to troubleshoot the user experience is low. While there are numerous tools that manage and monitor specific aspects of the cloud, an end-to-end tool does not exist. For such a tool to be created, standards need to be agreed to and implemented for all aspects of the cloud.

Because of the size of the virtual machines, the network infrastructure may require upgrading.

Appendix 8. Virtual Desktop Infrastructure Draft Use Case

Business Use Case

Created By:	Robert Patt-Corner	Last Updated By:	Robert Patt-Corner
Date Created:	1/14/2011	Date Last Updated:	1/14/2011
Version	0.9	Changes	Initial

1. Use Case Identification

1.1. Use Case Name

Shared GeoSpatial Platform-as-a Service

1.2. Agency

Federal Geographic Data Committee, with individual project contributions from member agencies including USGS (U.S. Geologic Survey), NOAA (National Oceanic and Atmospheric Administration), Bureau of the Census, EPA (Environmental Protection Agency), USDA (Department of Agriculture), DOI (Department of the Interior) with interest from DHS (Department of Homeland Security)

1.3. Model Matrix

Identify which intersections of the service/deployment matrix the use case addresses.

		Cloud Service Models		
		Software as a Service (SaaS)	Platform as a Service (PaaS)	Infrastructure as a Service (IaaS)
Deployment Models	Private			
	Community	X	X	(X – involved)
	Public	X	X	(X – involved)
	Hybrid			

1.4. Created By

Robert Patt-Corner, Federal Cloud Computing Initiative

1.5. Date Created

January 14, 2011

1.6. Last Updated By

Robert Patt-Corner, January 14, 2011

1.7. Date Last Updated

Robert Patt-Corner, January 14, 2011

2. Background

The Federal Geographic Data Committee with the GSA cloud computing Program Management Office are operating the GeoCloud project on behalf of a wide range of Federal Agencies to explore the impact and possibilities of a geospatially-oriented cloud. The initiative seeks to define and investigate cloud savings, best practices and lessons learned by migrating, benchmarking and operating a set of 10 existing public-access geospatial projects from 6 currently participating agencies – USGS (U.S. Geologic Survey), NOAA (National Oceanic and Atmospheric Administration), Bureau of the Census, EPA (Environmental Protection Agency), USDA (Department of Agriculture), DOI (Department of the Interior) with interest from DHS (Department of Homeland Security).

The overall plan is to define, construct and maintain a set of common geospatial platforms to support the projects, using a joint agency platform model. Once platforms are in place and under maintenance, each project team will evaluate their application on its matching platform, documenting the steps needed to ensure security and performance, and tracking lessons learned along the way. To date two platforms have been defined, one has been hardened and constructed and is operating on Amazon's AWS public cloud. The project teams are beginning their exploration and sandbox phase to discover and document the processes needed to maintain these existing applications in the cloud.

Some agency geospatial applications over and above the 10 projects targeted to the public cloud have either data storage or processing needs that appear to make them more cost effective in a community cloud setting. As a subsidiary use case, these applications will be piloted on similar shared platforms in a community facility housed in the U.S. Geologic Survey.

3. Definitions

FGDC: Federal Geographic Data Committee is an interagency committee that promotes the coordinated development, use, sharing, and dissemination of geospatial data on a national basis. http://www.fgdc.gov/

GIS: Geographic Information System -- any system that captures, stores, analyzes, manages, and presents data that are linked to location(s). In the simplest terms, GIS is the merging of cartography, statistical analysis, and database technology. http://en.wikipedia.org/wiki/Geographic_information_system

ESRI: Vendor of Geographic Information Systems (GIS) widely used in Federal Government and Industry, and focus of one of the two prototype GeoSpatial platforms.

GeoCloud: Name of the FGDC project exploring Geospatial Cloud Platform for contributed member geospatial projects

GeoSpatial Cloud Platform: Cloud-based Platform as a Service operated by the FGDC, and tailored to support deployment of a wide range of pre-identified GeoSpatial applications.

GeoNet: A project running on the Geospatial Cloud Platform that provides both standalone Geospatial services and additional Geospatial SOA capabilities that can be leveraged by layered applications.

4. Primary Actors

Platform Managers: Platform Managers are responsible for the architecture and design of the shared Geospatial Cloud platform implementations. Platform managers function in the role of Platform Providers with respect to the Draft NIST Definition of Cloud Computing service model. Their scope of activity is the Geospatial Platforms.

Platform Developers: Platform Developers are responsible for the creation, composition and maintenance of the shared Geospatial Cloud platform implementations. Platform Developers function in the role of Platform Providers with respect to the Draft NIST Definition of Cloud Computing service model. Their scope of activity is the Geospatial Platforms.

Application Deployers: Application Deployers are responsible for the deployment of each project's applications on an appropriate GeoSpatial Cloud Platform implementation. Deployers function in the role of Platform Consumers with respect to the Draft NIST Definition of Cloud Computing service model. Their scope of activity is the individual GeoSpatial project whose applications are deployed to the platform.

Application Managers: Application Managers are responsible for the operation, monitoring and maintenance of each project's applications on an appropriate GeoSpatial Cloud Platform implementation. Application Managers function in the role of Platform Consumers with respect to the Draft NIST Definition of Cloud Computing service model. Their scope of activity is the individual GeoSpatial project whose applications are running on the platform.

Note that application developers are not present as a role because the initiative currently draws from a pool of already-existing applications

5. Business Goal

The goals of the FGDC GeoCloud initiative are to establish, benchmark, operate, document and certify a common Geospatial Cloud Platform using shared geospatial Cloud "Platform as a Service" images and common procedures. These images and the security certification are intended to be shared by agencies for quick provisioning of geospatial services and data with a significant reduction in C&A (A&A) workloads

6. Service model

Platform as a Service (PaaS) was selected in order to gain economies of scale in deploying multiple geospatial applications to the cloud. Analysis showed that the set of candidate applications had significant common platform requirements, and that extracting, maintaining and providing those requirements as a shared platform offered substantial cost, operational and logistical advantages over deploying each application uniquely in an IaaS model.

IaaS was able to meet the requirements of the candidate applications, but at the cost of diverging implementations and consequent increased expenses and decreased efficiency and cross-project leverage.

SaaS was inappropriate because of the divergent business purposes of the 10 candidate applications, however the solution has some service-oriented SaaS elements in the web services provided to any project by some of the applications, including GeoNet and ERDDAP.

7. Deployment model

Public cloud was selected as the initial target for all GeoCloud applications because of the public-facing unrestricted data characteristics of the target applications, the expected cost savings and the wide availability of suitable public infrastructure. Subsequent analysis indicated that a subset of candidate applications over and above the 10 current public cloud candidates had processing and data storage needs that were more cost effective to meet in a community cloud, and these are proceeding on a parallel community cloud track. Revisions in public cloud pricing may cause the project to periodically reevaluate and benchmark these applications' suitability for migration to public cloud.

8. Necessary Conditions

Identify the conditions that must be met for the system or service to be implemented successfully.

8.1. Security

Infrastructure must be accredited at a low FISMA level of security, with moderate preferred.

Interconnection Security Agreements. (ISA) must be implemented, ideally for the platform as a whole, but if that is not possible for policy reasons, on a template basis for each agency or project in the initiative. Finding the optimal means to leverage accreditation agreements is a key deliverable of the project.

8.2. Interoperability

Initially the GeoCloud is targeting a single public infrastructure using Amazon AWS. Eventually as the project moves beyond early stages to full operation it will ideally operate platforms on any of the IaaS Awardees for the Infrastructure-as-a-Service GSA Blanket Purchase Agreement. In addition, the initial public platforms will benefit from being able to operate on the community cloud infrastructure, and visa-versa.

To that end, the following interoperability requirements are needed:

- ☐ Authentication and identity management interoperability will be required so that users of multiple target clouds can maintain consistent identity and role-based access across multiple cloud implementations.
- ☐ Virtual machine management interoperability will be required so that platforms running in multiple cloud implementations can be stopped, started, terminated and otherwise operated through a consistent interface.
- ☐ Billing and reporting interoperability will be helpful to allow for meaningful comparisons of costs and benefits across multiple cloud implementations.

8.3. Portability

Static virtual machine portability is required so that the maintained platform images can be freely migrated between cloud implementations without the need for parallel development or maintenance. Dynamic VM portability, where running machines are migrated in flight is not required.

8.4. Other

Categorize and describe any other business needs not addressed above.

9. Priorities and Risks

The GeoCloud is a high priority but low risk project for FGDC, as it is exploratory in nature and intended to provide justification for higher reward and risk operational projects based on its findings and lessons learned. Because the main focus is information gathering.

10. Essential Characteristics

Describe how the system meets the five essential characteristics of a cloud computing system along with the benefits provided by each of these characteristics.

☐ On-demand self service

- Platform images are able to be created and modified on demand and as needed by Platform Deployers and Managers
- Platform images are able tbe started as virtual machines by any of the four principal actors; charges are limited those instances actually running at a point in time
- GeoSpatial Applications are deployed on platform instances as required by Application Deployers and can be started and stopped on demand by Platform Managers.

☐ Broad network access

- The capabilities of the GeoCloud are freely available to authorized roles over the public internet, or alternately over virtual private network at the discretion of the Application Manager, based purely on security considerations.

☐ Resource pooling

- At the Platform level, the applications making up the projects in GeoCloud make use of a smaller set of target Geospatial platforms than there are applications. Currently twtarget platforms serve the needs of 10 identified applications.
- At the infrastructure level, Platform Managers and Developers draw on resources from a pool of resources spread across two nationwide availability zones, without reference to particular hardware or physical location.

☐ Rapid elasticity

- Instances running the platforms can be started, stopped and scaled at will, including scaling by rule based on load-based parameters
- Storage for the applications is freely expandable and contractable on demand.

☐ Measured service

- Service is metered on an hourly basis and billed monthly based on actual resources consumed. Some cost advantages are expected from committing to a minimum quantity of resource usage, but this is purely a cost-reduction measure and is not essential to operation
- Service usage can be accessed and analyzed at any time on a near-real-time basis.

11. Normal Flow

Use Case 1: Create, Compose and Harden Target Platforms

Actors: Platform Providers (Platform Managers and Developers)

Goal: Provide a working initial Geospatial Platform release

Brief Description: A noncloud operating system is hardened to security specifications

The operating system is ported to the cloud environment as a private FGDC offering.

Requisite platform enablers are installed, configured, scripted, hardened and saved as a virtual machine image to form a point release of a flavor of the Geospatial platform.

Platform developers document lessons learned, best practices and installation operations as part of the deliverables of the project.

The platform release is advertised to consumers of the GeoSpatial Platform through the project portal.

Use Case 2: Maintain and Monitor Target Platforms

Actors: Platform Providers (Platform Managers and Developers)

Brief Description: Relevant information sources are monitored for security update and platform enabler enhancement information

Platform clients are engaged in a regular process to solicit desired changes and enhancements to the platform based on new project needs, new projects added or current deficiencies.

On a regular basis, or as an emergency activity in cases of high risk events, the platform is updated to reflect security and/or functionality enhancements.

Release notes and lessons learned for each point release are documented.

Changes are advertised to Platform Consumers, with a time window for migration

Use Case 3: Develop Project Application Deployment Packages

Actors: Platform Consumers (Application Managers and Deployers)

Brief Description: Application deployers access a version of the appropriate Geospatial Platform and install their application on the platform, noting all activities, dependencies and issues

Application deployers inform Project Developers of any platform deficiencies through the project portal. Deployers may wait on a new project version or include enhancements temporarily in their deploy procedure.

Application deployers analyze the install activities and create deployment scripts, which along with the code base and data of the application consititute a deployment package, typically one per application tier.

Application deployers iteratively test, benchmark and refine the deployment packages to an acceptable level of performance and ease of use.

Application deployers document lessons learned, best practices and script and package operation as part of the deliverables of the project

Use Case 4: Deploy Project Applications

Actors: Platform Consumers (Application Managers and Deployers)

Brief Description: Application deployers execute their application's deployment scripts, which deploy and start the applications deployment packages on the target cloud environment.

Use Case 5: Manage and Benchmark

Actors: Platform Consumers (Application Managers and Deployers)

Brief Description: Application deployers execute their application's deployment scripts, which deploy and start the applications deployment packages on the target cloud environment.

Application Managers operate, monitor, benchmark and document the performance and cost of the application as part of the deliverables of the project.

Use Case 6: Upgrade Application to Revised Platform

Actors: Platform Consumers (Application Managers and Deployers)

Brief Description: Application deployers execute their application's deployment scripts against the advertised revised platform in a controlled nonproduction environment.

Application deployers revise deployment packages and/or notify Platform Developers of issues or defects from the results of their tests.

On production of acceptable revised packages and/or modifications to the new image Application deployers start new instances of the application tiers on the upgraded platform and adjust domain name entries to shift user load from old to new platform.

12. Frequency of Use

Use Case 1: Create, Compose and Harden Target Platforms

Frequency: Once per target platform

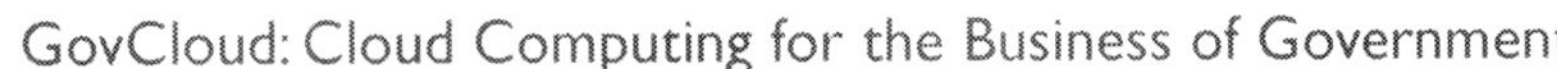

Use Case 2: Maintain and Monitor Target Platforms

Frequency: Regular, minor changes expected Monthly, major changes biannual

Use Case 3: Develop Project Application Deployment Packages

Frequency: Once per project application

Use Case 4: Deploy Project Applications

Frequency: Periodic, corresponding to application or platform updates

Use Case 5: Manage and Benchmark

Frequency: Ongoing and continuous

Use Case 6: Upgrade Application to Revised Platform

Frequency: Periodic, corresponding to application or platform updates

13. Special Requirements

Identify any additional requirements for the use case that need to be addressed. Examples include performance or availability requirements.

14. Notes and Issues

List any additional comments about this use case or any remaining open issues.

CPSIA information can be obtained at www.ICGtesting.com
224212LV00001B/4/P

9 780983 236139